Smart Sensors

This reference text comprehensively discusses micro-electromechanical systems and nanoelectromechanical systems-based design of smart sensors, fabrication techniques for smart sensors, and potential application areas. It covers applications of smart sensors in diverse areas including medical, agricultural, space, automobiles, manufacturing, security, and surveillance.

This book

- discusses the design parameters of micro-electromechanical systems and nanoelectromechanical systems-based smart sensors;
- covers smart sensors for the conditioning and monitoring of electrical machines, robotic systems, and electric vehicles;
- highlights the importance of using smart sensors in localization, navigation, and mapping;
- explains efficient mobile ad hoc networks and smart sensor technologies for the Internet of Things applications;
- illustrates how smart sensors can be used in different applications like biomedical systems, home automation and other related areas.

It is primarily written for senior undergraduates, graduate students, and academic researchers in the fields of electrical engineering, electronics and communications engineering, sensor technology, nanoscience, and nanotechnology.

Smart Sensors
Design, Challenges, and Applications

Edited by

Manoj Kumar Shukla, Praveen Kumar Malik,
Anuj Jain, and Neeraj Kumar Mishra

CRC Press
Taylor & Francis Group
Boca Raton London New York

CRC Press is an imprint of the
Taylor & Francis Group, an **informa** business

MATLAB® and Simulink® are trademarks of The MathWorks, Inc. and are used with permission. The MathWorks does not warrant the accuracy of the text or exercises in this book. This book's use or discussion of MATLAB® and Simulink® software or related products does not constitute endorsement or sponsorship by The MathWorks of a particular pedagogical approach or particular use of the MATLAB® and Simulink® software.

Designed cover image: shutterstock

First edition published 2025
by CRC Press
2385 NW Executive Center Drive, Suite 320, Boca Raton FL 33431

and by CRC Press
4 Park Square, Milton Park, Abingdon, Oxon, OX14 4RN

CRC Press is an imprint of Taylor & Francis Group, LLC

ISBN: 978-1-032-74663-0 (hbk)
ISBN: 978-1-041-06125-0 (pbk)
ISBN: 978-1-003-63388-4 (ebk)

DOI: 10.1201/9781003633884

Typeset in Sabon
by codeMantra

Contents

Preface

The rapid advancement of technology in recent decades has fundamentally transformed various industries, leading to the emergence of smarter and more efficient systems. At the heart of these innovations are smart sensors—devices that not only measure physical quantities but also process and communicate the data, enabling intelligent decision-making in real time. This book aims to provide a comprehensive exploration of this dynamic field, offering insights into the intricate design processes, the multifaceted challenges faced, and the wide-ranging applications of smart sensors.

Smart sensors have become integral to numerous domains, including healthcare, environmental monitoring, industrial automation, automotive systems, and the Internet of Things. They are the building blocks of intelligent systems that enhance the quality of life, improve operational efficiency, and contribute to sustainable practices. However, the journey from concept to deployment is fraught with challenges. Designing smart sensors requires a deep understanding of multidisciplinary fields such as electronics, materials science, signal processing, and data analytics. Furthermore, ensuring reliability, accuracy, energy efficiency, and security in diverse and often harsh environments presents significant hurdles.

This book is structured to guide the reader through these complexities. The first section delves into the fundamental principles and design strategies of smart sensors, emphasizing the integration of sensing, processing, and communication functions. The second section addresses the challenges encountered in the development and implementation of smart sensors, including technical, logistical, and ethical considerations. Finally, the third section showcases a variety of applications, illustrating how smart sensors are being employed to address real-world problems and innovate across sectors.

We hope that this book serves as a valuable resource for students, researchers, engineers, and professionals who are engaged in the field of smart sensors. By shedding light on both the theoretical and practical aspects of smart sensors, we aim to inspire further exploration and innovation in this exciting and rapidly evolving domain.

Acknowledgements

The completion of this book would not have been possible without the collaborative efforts, guidance, and support of many individuals and institutions. As editors, we would like to express our deepest gratitude to those who have contributed to this project in various capacities.

First and foremost, we extend our sincere thanks to Symbiosis Institute of Technology, Symbiosis International University, Pune, Maharashtra; Lovely Professional University, Phagwara, Punjab; and N.I.M.S University, Jaipur, Rajasthan. These esteemed institutions provided us with the academic environment and resources necessary to bring this project to fruition. Their commitment to fostering research and innovation in the field of smart sensor technology has been instrumental in shaping the content of this book.

We are particularly indebted to our contributing authors, who have shared their expertise and insights through their chapters. Their dedication to advancing knowledge in this field is evident in the depth and quality of their contributions. Without their hard work, this book would not have been possible. We also appreciate the peer reviewers who provided valuable feedback and helped ensure the academic rigor and relevance of the content.

We would like to express our heartfelt gratitude to our colleagues and students at our respective institutions. Their encouragement, support, and willingness to engage in meaningful discussions have enriched our understanding and driven us to explore new perspectives. Special thanks are due to the research teams and laboratory staff who assisted in various aspects of the project, from experimental setups to data analysis.

We are also grateful to our families for their unwavering support and patience throughout the editing process. Their understanding and encouragement have been a source of strength during the long hours of work required to complete this book.

Lastly, we extend our appreciation to the publisher and editorial team for their professionalism, guidance, and support throughout the publication process. Their dedication to producing a high-quality book has been invaluable.

It is our hope that this book will serve as a valuable resource for students, researchers, and professionals working in the field of smart sensors, and that it will inspire further innovation and exploration in this rapidly evolving domain.

Manoj Kumar Shukla
Praveen Kumar Malik
Anuj Jain
Neeraj Kumar Mishra

About the Editors

Manoj Kumar Shukla is currently working as an Assistant Professor in the Department of Robotics and Automation at Symbiosis Institute of Technology, Symbiosis International (Deemed University), Pune. He completed MTech and PhD in Electrical Engineering from NIT, Hamirpur (Himachal Pradesh), India, in 2012 and 2019, respectively. His research interests mainly include control systems, robotics, artificial intelligence, smart sensors, IoT and automation. He has guided many BTech projects and MTech dissertations. He has 10 years of teaching and research experience. He is a reviewer of various reputed journals and has attended various international conferences. He has published research papers in various reputed journals and conferences.

Praveen Kumar Malik is a Professor in the School of Electronics and Electrical Engineering, Lovely Professional University, Phagwara, Punjab, India. He received his PhD with a specialization in Wireless Communication and Antenna Design. He has authored or coauthored more than 150 technical research papers published in leading journals and conferences from publishers of international repute. Some of his research findings are published in top-cited journals. He has also published ten edited/ authored books with international publishers. He has guided many students leading to ME/MTech and is guiding students leading to PhD. He is an Associate Editor of different journals. His current research interests include microstrip antenna design, MIMO, vehicular communication, and IoT. He was invited as Guest Editor/Editorial Board Member for many international journals, as Keynote Speaker in many international conferences held in Asia, and as Program Chair, Publications Chair, Publicity Chair, and Session Chair in many international conferences. He has been granted two design patents and a few are in pipeline.

Anuj Jain is a dedicated academician and researcher currently working at Lovely Professional University in Punjab, India. Dr. Jain is a distinguished academician and researcher with a strong foundation in Electronics and Communication Engineering. Through his teaching career and research contributions, he has made significant strides in the domains of network security, IoT, and artificial intelligence. His commitment to knowledge dissemination and his ability to inspire and guide students have earned him respect within the academic community. Dr. Jain's work continues to shape the future of research and education, making him a valuable asset in the field.

Neeraj Kumar Mishra is currently working as an Assistant Professor in the Department of Electrical Engineering at NIT Hamirpur, HP, India. He earned his BE in Electrical Engineering and ME in Measurement & Control from the Madhav Institute of Technology and Science, an autonomous institute under the MP State Government, India, in 2007 and 2011, respectively. He received his PhD from the National Institute of Technology in Hamirpur, India, in 2022. As a distinguished academician and researcher, Dr. Mishra has made a significant impact at various prestigious NITs, including NIT Hamirpur and NIT Delhi. With a passionate commitment to teaching and an unyielding dedication to groundbreaking research, Dr. Mishra has notably advanced the field of six-phase doubly fed induction generators throughout his illustrious career. His extensive expertise spans multiple domains, including power electronics, machines, drives, wind energy generation systems, electric vehicles, conventional and nonconventional energy generation systems, artificial intelligence, machine learning, and artificial immune systems.

Contributors

Manoj Kumar Bandil
Department of Electrical Engineering
ITM University
Gwalior, India

Lalit Chaudhari
Department of Instrumentation Engineering
Dr. D. Y. Patil Institute of Technology
Savitribai Phule Pune University
Pune, Maharashtra, India

Shubhangi Deokar
Department of Computer Science and Engineering, Symbiosis Institute of Technology
Symbiosis International (Deemed University), (SIU)
Pune, Maharashtra, India

Jayprakash Giri
Department of Electrical Engineering
NIMS University
Jaipur, Rajasthan, India

Rolly Gupta
Department of Computer Science Engineering
SRMIST
Delhi NCR, Ghaziabad, India

Ashwini Jatti
Department of Instrumentation Engineering
Dr. D. Y. Patil Institute of Technology
Savitribai Phule Pune University
Pune, Maharashtra, India

Anita Jaswal
Department of Agronomy, School of Agriculture
Lovely Professional University
Phagwara, Punjab, India

Anuradha Jayaraman
Department of Environmental Science
Nims Institute of Allied Medical Science and Technology
Nims University Rajasthan
Jaipur, India

Kabita Kumari
Department of Instrumentation and Control Engineering
Dr. B. R. Ambedkar National Institute of Technology
Jalandhar, Punjab, India

Samir Kumar
Department of Electronics and Information Engineering
Korea University
Sejong, Republic of Korea

Sanjeev Kumar
Biomedical Instrumentation Unit
CSIR-Central Scientific Instruments Organisation (CSIR-CSIO)
Chandigarh, India

Arshi Naim
Department of Business Management
European Global Institute of Innovation and Technology
Malta

Sharvan Kumar Pahuja
Department of Instrumentation and Control Engineering
Dr. B. R. Ambedkar National Institute of Technology
Jalandhar, Punjab, India

Prantick Patra
Department of Agronomy, School of Agriculture
Lovely Professional University
Phagwara, Punjab, India

Sanjeevi Ramakrishnan
Department of Environmental Science
Nims Institute of Allied Medical Science and Technology
Nims University Rajasthan
Jaipur, India

Parul Raturi
Department of Physics
Omkaranand Sarswati Government Degree College
Devpryag, Uttarakhand, India

Lalit Kumar Sagar
Department of Computer Science Engineering
SRMIST
Delhi NCR, Ghaziabad, India

Shimpy Sarkar
Department of Entomology, School of Agriculture
Lovely Professional University
Phagwara, Punjab, India

Prashantkumar B. Sathvara
Nims Institute of Allied Medical Science and Technology
Nims University
Jaipur, Rajasthan, India

Javed Sayyad
Department of Robotics and Automation, Symbiosis Institute of Technology
Symbiosis International (Deemed University) University (SIU)
Pune, Maharashtra, India

Bhavya Shukla
Department of Computer Science & Engineering
UIT
Shivpuri, Madhya Pradesh, India

Manoj Kumar Shukla
Symbiosis Institute of Technology
Symbiosis International (Deemed University)
Pune, India

Arshdeep Singh
Department of Agronomy, School of Agriculture
Lovely Professional University
Phagwara, Punjab, India

Maninder Singh
Department of Agronomy, School of Agriculture
Lovely Professional University
Phagwara, Punjab, India

Sandeep Tripathi
Nims Institute of Allied Medical Science and Technology
Nims University Rajasthan
Jaipur, India

Yogesh Kumar Verma
School of Electronics and Electrical Engineering
Lovely Professional University
Jalandhar, India

Anup Vibhute
Department of Instrumentation Engineering
Dr. D. Y. Patil Institute of Technology
Savitribai Phule Pune University
Pune, Maharashtra, India

Arun Kumar Wadhwani
Department of Electrical Engineering
MITS
Gwalior, India

Harsh Wadhwani
Department of Computer Science & Engineering
UIT
Shivpuri, India

Chapter 1

Multifunctional SERS substrates for smart sensing

Principles, applications, and emerging trends

Parul Raturi and Samir Kumar

1.1 INTRODUCTION

Surface-enhanced Raman spectroscopy (SERS) is an analytical technique that offers significant advantages in terms of sensitivity and limit of detection compared to conventional Raman spectroscopy.[1–4] By amplifying Raman signals from molecules adsorbed on metal surfaces, SERS enables the detection of even single molecules.[5,6] The underlying principle of SERS is the electromagnetic enhancement (EM) resulting from the interaction of molecules with plasmonic nanostructures, which can reach extremely high values of up to 10^{10}.[7] This enhancement occurs through two main contributions: local field enhancement and re-radiation enhancement due to the presence of the metallic structure.[8]

The majority of research in the field of SERS has focused on enhancing the electromagnetic properties of SERS substrates by optimizing the geometry of the nanostructures or the gaps/pores between them.[9–11] It has numerous applications in chemical and biological sensing, environmental monitoring, and material science.[12–15] In particular, SERS is well suited for ultrasensitive detection of trace amounts of analytes, such as drugs, explosives, and pathogens.[16–19] The capability of SERS to identify individual molecules has opened up new possibilities for detecting biomolecules like proteins and DNA without requiring labeling, even in complex biological environments.[20–23]

However, conventional SERS substrates fabricated using top-down lithography techniques or bottom-up chemical methods often suffer from limitations such as poor reproducibility and lack of renewability.[24–30] To overcome these limitations and expand the potential applications of SERS in smart sensing, there has been a growing interest in designing multifunctional SERS substrates that integrate additional functionalities beyond plasmonic enhancement.[31–36]

Multifunctional SERS substrates incorporate active components and tunable properties, enabling dynamic control during the sensing process.[37–43] These substrates offer advantages such as renewability, multiplexed and multimodal detection, and integrating functional materials

DOI: 10.1201/9781003633884-1

for dynamic tuning or analyte immobilization.[44–49] For example, incorporating functional materials such as phase-change materials,[50] nanowires,[37] field-effect transistors,[51] and hydrogels enables dynamic tuning of plasmonic enhancement or analyte immobilization.[52,53] By incorporating renewable functionalities, these substrates can be regenerated and reused when analytes are removed or degraded.[54] Thus, multifunctional SERS platforms have the potential to realize real-time, multi-analyte, and responsive SERS-based sensors for a wide range of sensing and biomedical applications.

This chapter presents the recent advancements and upcoming trends in multifunctional SERS substrates for smart sensing purposes. We will start by examining the basic principles of SERS signal amplification and different methods for producing plasmonic nanomaterials. Following that, we will investigate the various applications that have gained advantages from multifunctional SERS sensing platforms, which include detecting chemicals and biomolecules, medical diagnostics, monitoring the environment, and evaluating food safety. Finally, we will look into emerging trends such as incorporating machine learning (ML), flexible and wearable substrates, and in vivo sensing.

1.2 FUNDAMENTAL PRINCIPLES OF SERS

SERS offers remarkable sensitivity and selectivity in detecting molecules at extremely low concentrations.[55] It amplifies Raman scattering signals using plasmonic nanostructures, such as metal nanoparticles and nanostructured surfaces. Unlike traditional Raman spectroscopy, SERS can detect even a single molecule adsorbed onto these nanostructures.

The enhancement observed in SERS is primarily attributed to the nanostructures' localized surface plasmon resonances (LSPRs). LSPRs amplify the electromagnetic field and enhance the Raman scattering cross-section of the scattering molecules (see Figure 1.1).[56] Several factors influence the SERS enhancement, including the plasmonic nanostructures' size, shape, and composition.[57–60] The size and shape of the nanostructures determine the LSPR frequency, which determines the wavelength of light most strongly enhanced.[2,8] By altering the size or shape of the nanostructures, the LSPR spectra can be conveniently adjusted across a wide range of wavelengths, from near-ultraviolet to visible spectrum and even into mid-infrared.[61] Moreover, the composition of the nanostructures also affects the enhancement, as different metals possess distinct LSPR properties.[62,63] Katyal discusses the LSPR properties and field enhancement of different metals with different nanostructures, indicating that the composition of the nanostructures can affect the SERS enhancement.[63] The surface chemistry of the nanostructures also plays a role in the strength of the interaction

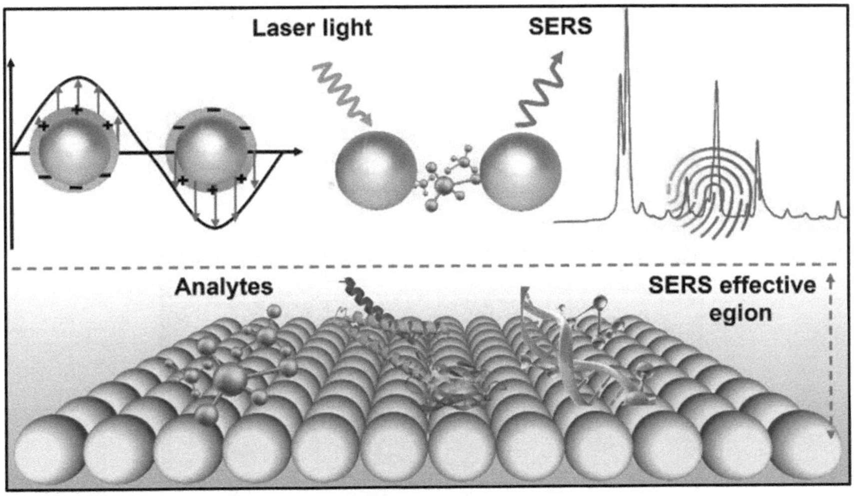

Figure 1.1 A visual representation showing the mechanism of surface-enhanced Raman scattering (SERS). (Reproduced with permission from Ref. [64]. Copyright 2022 Taylor & Francis.)

between the molecules and the nanostructures, further influencing the SERS enhancement.

Additionally, the surrounding environment of plasmonic nanoparticles, such as the substrate thickness and the thickness and dielectric parameters of the layer on the nanoparticle's surface, can be modified to control the position of the resonance wavelength.[65] This interaction is crucial for SERS enhancement, allowing the LSPRs to amplify the Raman scattering signals. Furthermore, the distance between molecules and nanostructures is critical for SERS enhancement, with the strongest enhancement observed when the molecules are closer to the nanostructures, typically at a distance of 1–30 nm (Figure 1.2). The distance dependence of SERS can be approximated using the equation:

$$I = \left(1 + \frac{r}{a}\right)^{-10}, \tag{1.1}$$

where I is the intensity of the Raman mode, a is the average size of the field-enhancing features on the surface, and r is the distance from the surface to the adsorbate.[66,67] By carefully controlling these factors, it is possible to optimize SERS enhancement and achieve highly sensitive and selective detection of molecules.

Understanding the fundamental principles and mechanisms underlying SERS enhancement is crucial for comprehensively grasping this technique and harnessing its potential in various applications. The interplay between

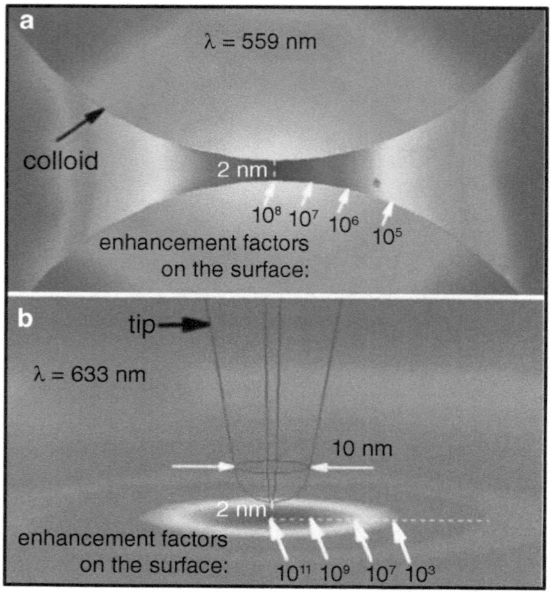

Figure 1.2 (a) Distance dependence of the enhancement factor distribution in the hotspot between two gold nanoparticles. (b) Distance dependences of the tip-enhanced Raman scattering enhancement factor for a gold tip positioned 2 nm above a gold surface in the air. (Reproduced with permission from Ref. [68]. Copyright 1999 Royal Society of Chemistry.)

plasmonic nanostructures, size, shape, composition, surface chemistry, surrounding environment, and distance between molecules and nanostructures contribute to the overall SERS enhancement. By studying and manipulating these factors, we can unlock the full potential of SERS for advanced sensing and detection applications. In the following section, we will briefly discuss the key factors that influence SERS enhancement.

1.2.1 Plasmonic nanostructures

The SERS enhancement relies on the intricate interplay between light and plasmonic nanostructures, a phenomenon extensively explored in scientific literature. Noble metal nanoparticles, particularly gold and silver, with exceptional plasmonic properties, are commonly employed in SERS. These nanoparticles exhibit LSPRs, which result from the collective oscillation of conduction electrons in response to incident light. LSPRs generate intense localized electromagnetic fields near the nanoparticle surface, significantly amplifying the interaction between light and molecules (Figure 1.3).[69,70]

There are numerous studies on the significance of nanostructure size, shape, and arrangement in tailoring the SERS enhancement effect.[71–77]

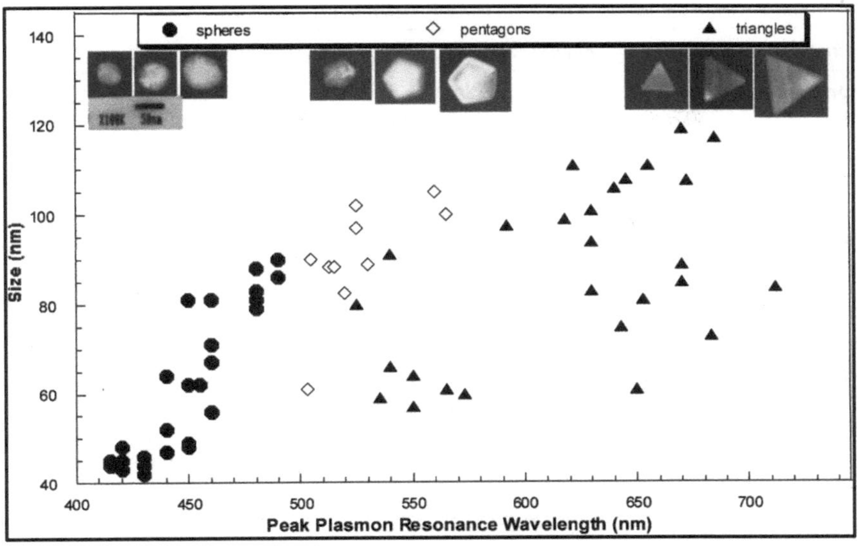

Figure 1.3 Relationship between nanoparticle lateral size and peak plasmon wavelength for various silver nanoparticles of different shapes. (Reproduced with permission from Ref. [83]. Copyright 2002 AIP Publishing.)

Nanostructures, such as nanospheres, nanorods, and nanoplates, possess distinct plasmonic characteristics that can be precisely tuned to optimize SERS performance. Nanospheres, for instance, scatter light isotropically, resulting in a relatively uniform enhancement across their surface.[28,78–81] In contrast, nanorods exhibit anisotropic plasmonic properties, with the enhancement localized at their tips, offering magnified sensitivity in specific directions. Nanoplates, with their two-dimensional (2D) geometry, provide additional degrees of freedom for plasmonic tuning, enabling precise control over the enhancement effect.[81,82]

Furthermore, the arrangement and organization of these nanostructures play a crucial role in collective plasmonic coupling, further enhancing the SERS effect. Assembling nanoparticles into ordered arrays or hierarchical architectures allows for synergistic enhancement effects.[84,85] Plasmonic coupling between adjacent nanoparticles creates "hotspots," where the electromagnetic fields are further intensified, leading to ultrasensitive SERS detection. The interplay between individual nanoparticle properties and collective plasmonic coupling offers a versatile platform for engineering SERS substrates with tailored enhancement characteristics.[86–89] Table 1.1 provides a comprehensive overview of the optical properties associated with diverse nanostructures. It is a foundational resource for researchers seeking to delve deeper into the optical characteristics of specific nanoparticle geometries. While the table offers a valuable starting point, it is essential

Table 1.1 Various optical properties related to different nanostructures[90]

Nanoparticle	*Material*	*References*	*Properties*
Nanosphere	Gold	[91]	Isotropic scattering, moderate enhancement. Attractive for quantitative optical characterization of single particle
Nanospheroid	Gold	[92]	Anisotropic scattering and enhancement. Transverse and longitudinal modes
Nanorod	Gold	[93,94]	Longitudinal and transverse resonances. Linear behavior with rod length. Modification of plasmon decaying rates
Nanodisk	Gold	[95]	Substrates to optimize biosensing. Useful as constituents for patch antennas
Nanoshell	Gold	[96]	Tunability due to shell thickness
Nanoring	Gold	[97]	Tunability due to ring thickness
Nanorice	Gold	[98]	Intense resonances, large field enhancements. Combines rod and shell properties
Nanocube	Silver	[99]	Appropriate for plasmon sensing
Nanostar	Gold	[100]	Intense hotspots at the nanostar tips
Nanoegg	Gold	[101]	Tunability, asymmetric system, Fano spectral profiles
Nanocup	Gold	[102]	Capability to bend light
Nanospiral	Gold	[103]	Complex response
Nanocrescent	Gold	[104]	Tunable narrow resonances in the mid-infrared with good figures of merit for sensing
Nanotriangle	Silver	[105]	Strong scattering in the red. Modes associated with the triangle edges
Nanoprism	Gold	[106]	Presence of sharp edges. Breaking of symmetry
Nanohole	Gold	[107,108]	Inverse symmetry. Babinet's principle. Similar trends for complementary electric and magnetic fields
L-shaped	Gold	[109]	No center of inversion symmetry. Strong polarization dependence. Presence of bulk-like plasmons

to recognize that the information presented is not exhaustive and may not encompass the earliest research conducted in the field.

1.2.2 Surface chemistry

Surface chemistry plays a critical role in SERS by influencing the enhancement of Raman signals. One key factor in determining the enhancement efficiency is the affinity between the substrate surface and the analyte molecules. Functionalizing the substrate surface with specific chemical groups can enhance the binding interactions between the substrate and analyte molecules, resulting in increased adsorption and favorable molecule orientation.[110] These modifications support stronger electromagnetic coupling, leading to enhanced Raman scattering.

Chemical functional groups on the surface significantly influence the electromagnetic field distribution in SERS by interacting with the plasmonic properties of the substrate. The traditional understanding has been that the electromagnetic mechanism, driven by plasmon-mediated enhancement of local electromagnetic fields, is the primary contributor to SERS, with chemical contributions playing a minor role. However, recent studies suggest a more complex interplay. For instance, a comprehensive resonant study of nanosphere lithography-based metallic substrates with covalently attached 4-mercaptobenzoic acid monolayers revealed that the SERS efficiency is not solely dependent on the electromagnetic mechanism but also on a chemical resonant contribution related to a metal-to-ligand electronic transition of the probe molecule. This synergy between plasmon modes and ligand-to-metal chemical resonance maximizes amplification when both mechanisms intersect.[111] Additionally, the density of hotspots, which are regions of intense electromagnetic fields, can be enhanced by functionalizing nanostructures such as carbon nanowalls with plasmonic nanoparticles. This functionalization increases the specific surface area and charge mobility, thereby contributing to SERS's electromagnetic and chemical mechanisms.[112] Furthermore, the choice of surface modifiers in synthesizing plasmonic nanomaterials can either enhance or inhibit SERS activity. Modifier-free synthetic approaches have been shown to produce surface-accessible nanomaterials with significantly enhanced plasmonic properties, allowing for better interaction with target molecules and revealing new surface chemistry phenomena previously overlooked.[113] Hence, the presence and nature of chemical functional groups on the surface are crucial in modulating the electromagnetic field distribution and overall SERS efficiency, necessitating a deep understanding of both electromagnetic and chemical mechanisms for optimal analytical applications. Therefore, it is crucial to understand and control the surface chemistry of SERS substrates to optimize the enhancement of Raman signals. By carefully designing the substrate surface with appropriate chemical functional groups, it is possible to selectively enhance the Raman signals of specific analytes, improve the sensitivity of SERS measurements, and enable the detection and identification of trace amounts of molecules.

1.2.3 Surface roughness and morphology

Surface morphology features are crucial in influencing the electromagnetic field distribution in SERS. The formation of sharp tips in gold nanostructures, induced by higher-energy laser pulses, significantly enhances the electromagnetic field due to LSPR, which is highly sensitive to the nanostructure's shape and size. This enhancement is critical for achieving high

SERS signals and an enhancement factor (EF) of up to 10^7, making it feasible for single-molecule detection.[114] Additionally, the ability to tune the plasmon absorption of nanostructured SERS substrates to match the wavelength of the exciting laser radiation (530–620 nm) further underscores the importance of surface morphology. Changes in the morphology of metal nanoparticles on silicon oxide surfaces can adjust the optical properties and, consequently, the electromagnetic field distribution, optimizing the SERS effect for specific applications.[115] Therefore, the precise control of surface morphology is essential for maximizing the electromagnetic field enhancement in SERS, which directly impacts the sensitivity and effectiveness of the technique in various sensing applications.[76]

Surface roughness plays a crucial role in enhancing the plasmonic properties of metal nanostructures, directly impacting the SERS EF. In the study by Feng et al., used a strong thiol ligand as a morphology-directing reagent for creating Au nanoparticles with tunable surface roughness. This tunability in surface morphology led to a continuous adjustment of the LSPR within the visible-NIR region, resulting in substantial SERS enhancement due to the specific surface roughness of the nanoparticles.[116] Similarly, the research by Ding et al. demonstrated that Au nanorods with adjustable surface roughness, achieved through a process involving PbS nanoparticles, exhibited improved transverse plasmon resonance due to plasmon coupling between the nanorods and surface-modified Au nanoparticles. This coupling created abundant hotspots from ultrasmall nanogaps, sharp tips, and uneven areas, which significantly enhanced the SERS detection of Rhodamine B (RhB). The study found that Au nanorods with the highest surface roughness had the highest Raman EF at 532 and 785 nm laser excitations.[117] Therefore, increasing surface roughness enhances the plasmonic properties of Au nanostructures, leading to a higher SERS EF. This relationship is attributed to creating more plasmonic hotspots and improved light absorption, which is critical for applications in SERS-based detection and other plasmonic-related fields.

1.2.4 Defects and impurities

The presence of defects and impurities on the substrate surface creates active sites for the adsorption and enhancement of Raman signals. Manipulating these features can significantly impact the surface chemistry of SERS substrates. Defects and impurities in semiconductor materials have emerged as pivotal factors in enhancing the performance of SERS substrates. Traditionally, noble metal-based substrates dominated SERS applications due to their high EFs. However, recent advancements have shown that semiconductor defect engineering can achieve comparable or superior SERS activity.[118] For instance, the introduction of intrinsic defects in 2D palladium di-selenide ($PdSe_2$) dendrites, such as selenium vacancies and line defects, has been shown to create hotspots that significantly enhance the

SERS signal, achieving an EF greater than 10^5 and enabling the detection of RhB at concentrations as low as 10^{-8} M.[119] Similarly, defect engineering in semiconductors like WO_{3-x} nanosheets, which are rich in oxygen vacancies, has been demonstrated to improve SERS sensitivity by facilitating effective photoinduced charge transfer resonance and providing more charge transfer pathways.[120] These defects modulate the surface state and electronic structure, enhancing the interaction between the substrate and analyte molecules. Furthermore, the use of silver nanoparticles (AgNPs) on activated carbon substrates has shown that the combination of plasmonic properties and defect sites can lead to stronger SERS signals compared to AgNPs alone, with detection limits for β-agonists like clenbuterol, ractopamine, and salbutamol reaching as low as 0.001 mg/L.[121] The multifaceted tunability of semiconductor materials through defect engineering meets specialized sensing demands and bridges the gap between conventional plasmonic and plasmon-free SERS substrates, making them highly attractive for various applications.[122]

1.2.5 Solvent environment

The solvent can alter the local field around the analyte molecules, critical for the EM mechanism.[123] For instance, a study using an atomistic electrodynamics-quantum mechanical method demonstrated that the presence of solvent molecules, such as water, can lead to an increased local field at the position of the analyte, thereby enhancing the SERS signal.[124] This enhancement is attributed to the solvent's ability to modify the dielectric environment around the nanoparticles, affecting the surface plasmon resonance and the local electromagnetic fields. Additionally, the solvent can influence the chemical enhancement (CHEM) mechanism by affecting the electronic transitions and charge transfer processes between the analyte and the metal surface. For example, changes in the solvent environment can lead to variations in the induced electron densities and charge flows near the molecule–metal interface.[125] Furthermore, the solvent environment can impact the stability and reproducibility of SERS substrates. For instance, storing SERS substrates in an oxygen-free environment, such as under vacuum or argon, has significantly reduced the spectral attenuation rates, thereby improving the time stability of the SERS signals.[126] Moreover, innovative substrate designs, such as the magnetically photonic chain-loading system, can dynamically modulate SERS signals by aligning magnetic photonic nano-chains in the solvent, creating more hotspots and enhancing the SERS effect.[127]

Overall, understanding and controlling the surface chemistry of SERS substrates is crucial for optimizing the enhancement of Raman signals and tailoring them for specific applications. By manipulating factors such as surface roughness, defects, impurities, and solvent environment, researchers can unlock the full potential of SERS.

1.2.6 SERS enhancements

EM and CHEM are the two primary mechanisms contributing to the amplification of Raman scattering in SERS. EM is mainly driven by plasmonic nanostructures, such as gold and silver nanoparticles, which create localized electromagnetic hotspots. These hotspots amplify the Raman scattering through a double enhancement mechanism.[128,129] Figure 1.4 illustrates SERS's twofold EM mechanism, depicting how local electromagnetic fields

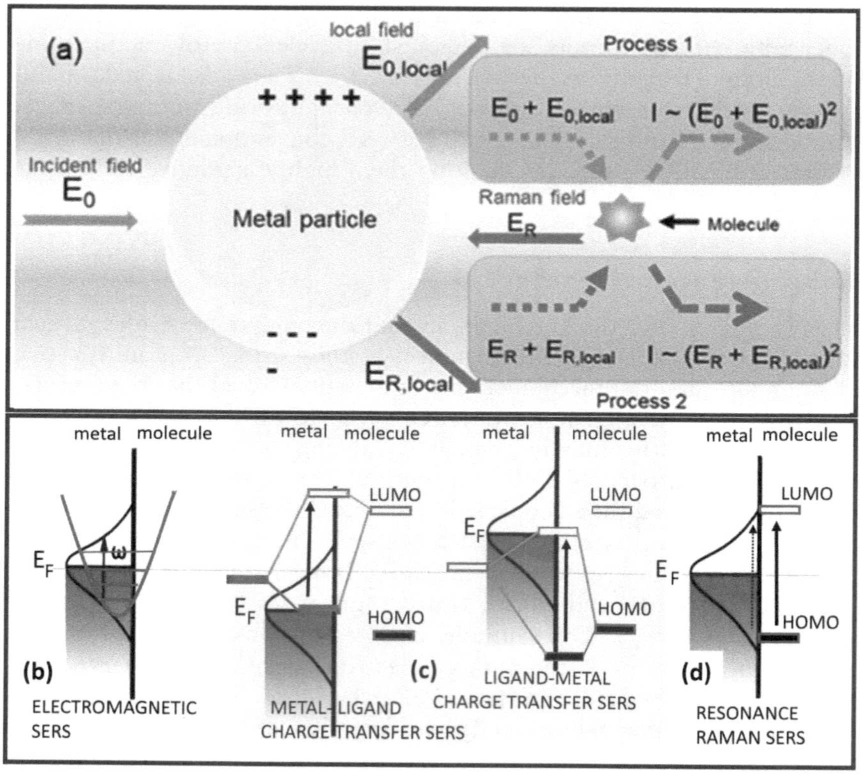

Figure 1.4 (a) Schematic illustrating the twofold electromagnetic enhancement mechanism in SERS.[8] The diagram depicts how electromagnetic fields are amplified near nanostructured surfaces, significantly enhancing the Raman signal from target molecules. (Reprinted (adapted) with permission from. Ref. [8]. Copyright 2020 American Chemical Society.) (b) Electromagnetic and chemical mechanisms for SERS Enhancement. Schematic illustration of the SERS enhancement mechanisms. (a) Electromagnetic (EM) enhancement arising from localized surface plasmon resonance excitation. (b) Chemical charge-transfer (CT) enhancement involves electron transfer between the analyte's highest occupied molecular orbital (HOMO) and the metal's Fermi level or vice versa, analogous to metal-to-ligand or ligand-to-metal charge transfer transitions. (c) Chemical Raman resonance (RR) enhancement due to interfacial HOMO-LUMO excitation. (Reproduced with permission from Ref. [131]. Copyright 2017 Elsevier.)

are amplified near nanostructured metal surfaces, typically gold or silver. Intense electric fields are generated around the nanoparticle surfaces when incident light excites localized surface plasmons. This increase in EM notably amplifies the Raman scattering signal of molecules situated in the proximity of these regions of high intensity called the "hotspots." The first enhancement originates from the increased local electric field, while the second stems from the re-radiation of the Raman-scattered light by the excited plasmons. Together, these effects dramatically improve the sensitivity of SERS.

CHEM mechanisms, in addition to physical effects like EM, significantly contribute to SERS. The formation of charge transfer complexes between the substrate surface and analyte molecules facilitates electron transfer, leading to changes in molecular polarizability and enhancing Raman signals. The nature of the chemical bonding between the substrate and analyte molecules also influences the enhancement. Covalent bonds, for example, promote more efficient charge transfer and greater enhancement than weaker interactions such as van der Waals forces.

Researchers have extensively studied the formation, distribution, and optimization of these hotspots to improve the performance of SERS. The seminal work by Nie and Emory emphasized the significance of nanoparticle shape, particularly sharp edges and tips, in hotspot generation.[56] Subsequent studies, such as those by Kneipp et al., further highlighted the importance of interparticle interactions in enhancing the Raman signals through nanoparticle aggregation.[5] These findings have provided valuable insights into the complex aspects of hotspot engineering.

Furthermore, researchers have incorporated advanced technologies like tip-enhanced Raman spectroscopy probes to advance hotspot design and fabrication.[130] The ongoing efforts in hotspot engineering, including developing novel nanoparticle assemblies and hierarchical structures, offer new avenues for optimizing hotspot distributions and the development of third-generation SERS hotspots. Studying hotspots' creation, distribution, and optimization will improve and progress SERS as an extremely sensitive spectroscopic method.

1.3 MULTIFUNCTIONALITY OF SERS SUBSTRATES

SERS substrates, engineered to amplify the Raman scattering signal of molecules adsorbed onto their surfaces, have become invaluable in various fields due to their multifunctionality beyond mere signal enhancement. Let us explore SERS substrates' diverse functions and applications in more detail.

1. **Signal Enhancement:** The primary function of SERS substrates is to enhance the Raman signal by exciting LSPR on metallic nanostructures, such as gold or silver. This enhancement can significantly increase the electromagnetic field at the surface, leading to a

remarkable increase in Raman signal intensity. This enhancement can lead to Raman signal increases by up to 10^6–10^{12} compared to normal Raman scattering.[132]

2. **Chemical Sensing:** SERS substrates find wide applications in chemical sensing. Their high sensitivity enables the detection of chemical species at trace levels. They are used to identify environmental pollutants, detect explosive materials, and monitor industrial processes. Functionalizing the substrate with selective binding agents, such as antibodies or aptamers, further enhances the specificity of SERS substrates for specific analytes.[133–135]
3. **Biological and Medical Applications:** In biomedical fields, SERS substrates play a critical role in detecting biomolecules like proteins, nucleic acids, and small metabolites. They are particularly useful in diagnostics for detecting disease markers at very low concentrations. SERS has been used to detect biomarkers for cancer, infectious diseases, and neurodegenerative conditions. Moreover, SERS substrates can be integrated into biosensors for real-time monitoring of physiological processes.[136,137]
4. **Catalysis:** Some SERS substrates possess catalytic properties and signal enhancement capabilities. Semiconductor and plasmonic nanostructures hybrid substrates can promote catalytic reactions, including photocatalytic reactions under light illumination. This makes SERS substrates valuable in applications such as pollutant degradation and synthetic chemistry.[44,137,138]
5. **Multimodal Sensing:** Advanced SERS substrates are designed for multimodal sensing, combining SERS with other analytical techniques. For example, integrating SERS with fluorescence or electrochemical sensing provides complementary information about the analyte, enhancing the overall analytical capability. This multifunctional approach enables more comprehensive analysis in complex matrices.[46,139,140]
6. **Optical Tweezers and Manipulation:** The strong electromagnetic fields around SERS-active sites can be utilized for optical tweezer applications, where small particles or biological cells are manipulated using laser beams. This capability allows studying single-molecule interactions and cellular processes at high spatial resolution.[141,142]
7. **Data Encryption and Anti-Counterfeiting:** SERS substrates are used in data encryption and anti-counterfeiting measures. The unique SERS spectra of substrates, especially when combined with specific markers or patterns, serve as optical barcodes or security tags. This helps authenticate products, secure documents, and prevent forgery.[143–145]
8. **Integration with Other Technologies:** SERS can be integrated with other well-established technologies, such as microfluidic technology, which enhances the detection of trace components and reduces

interference from complex samples. This integration highlights the potential of SERS substrates for rapid, trace-level biological and environmental analysis without labels.[146]

The versatility of SERS substrates goes well beyond conventional Raman signal amplification, encompassing various fields such as chemical detection, medical diagnosis, catalysis, multimodal sensing, optical control, security purposes, and others. Continuous progress in nanofabrication techniques further broadens the functionalities and uses of SERS substrates, confirming their significance in contemporary analytical science.

1.4 ENGINEERING MULTIFUNCTIONAL SERS SUBSTRATES

The core of multifunctional SERS substrates lies in the intricate engineering of hotspot distributions and densities across substrates of different dimensions, ranging from zero-dimensional (0D) to three-dimensional (3D) structures.

1.4.1 Zero-dimensional (0D) substrates

These substrates typically consist of isolated metal nanoparticles or nanocrystals are synthesized and functionalized to achieve tailored optical properties and specific molecular interactions.[147] These substrates enhance the Raman signal through LSPRs resulting from light interacting with the electrons on the surface of nanoparticles. The high surface-to-volume ratio of 0D nanoparticles maximizes the number of active sites for analyte binding and enhances the SERS signal. Additionally, the large surface area allows for functionalization with various capture agents, such as antibodies, aptamers, and peptides, enabling selective and specific detection of target molecules.[148–150] The optical properties of 0D SERS substrates can be precisely controlled by adjusting the nanoparticles' size, shape, and composition.[147] This tunability enables optimization of the LSPR wavelength for specific excitation sources and target molecules, enhancing the overall SERS performance. However, 0D SERS substrates face reproducibility, stability, and scalability challenges for practical applications.[151] Current research efforts are concentrated on enhancing the consistency and durability of nanoparticles, as well as investigating methods for large-scale production.[152–155]

1.4.2 One-dimensional (1D) SERS substrates

1D SERS substrates consist of highly ordered and aligned structures, such as nanowires, nanotubes, nanorods, or other elongated structures with dimensions less than 100 nm in two axes and significantly larger in the third that provide reproducible SERS signals.[155,156] These structures exhibit a high aspect ratio and large surface area and anisotropy in their SERS enhancement. The orientation of these nanostructures and the polarization of the incident light play a significant role in enhancing the SERS signal.[69,157–159] This property can be exploited to obtain additional information about the orientation of molecules adsorbed on the substrate surface, providing insights into molecular structure and dynamics. 1D SERS substrates can be fabricated using various techniques, including template-assisted growth, chemical vapor deposition, electrospinning, and glancing angle deposition.[160] These methods allow precise control over the size, shape, and composition of the nanostructures, enabling the tuning of SERS properties for specific applications. Challenges in this area include achieving uniform and reproducible SERS enhancement across the large substrate surface and integrating 1D SERS substrates with microfluidic devices and other sensing platforms (Figure 1.5).[161–163]

1.4.3 Two-dimensional (2D) SERS substrates

2D enhancing structures, such as arrays of closely spaced nanoparticles or patterned surfaces, provide a scalable approach for generating large hotspots with controlled geometries.[164] The tuning of interparticle distances and orientations in 2D substrates manipulates electromagnetic field distributions and enhances SERS signal intensities.[165] Metallic thin films and emerging materials like graphene, hexagonal boron nitride, and transition metal dichalcogenides are used in 2D substrates.[9,13,166,167] The high surface-to-volume ratio of 2D materials allows efficient analyte capture and signal amplification in SERS. Additionally, the atomically thin nature of 2D materials enables strong light–matter interactions, leading to enhanced Raman signals.[168] 2D materials are flexible and easily integrated into various substrates, including flexible and wearable devices. This versatility makes them suitable for various applications, such as point-of-care diagnostics, environmental monitoring, and chemical sensing. The discovery of 2D materials has significantly expanded the range of SERS applications, offering specificity to particular molecules and the ability to integrate benefits from both chemical and electromagnetic mechanisms. Graphene, the first 2D material, has showed the strong correlation between the EF in SERS and the proximity of chemical groups to the graphene surface.[169] Moreover, the integration of graphene with metallic nanostructures have improved SERS activity, expanding the scope of SERS applications.[170–172]

1.4.4 Three-dimensional (3D) SERS substrates

3D SERS substrates possess complex and hierarchical structures that maximize the SERS effect by creating many hotspots.[173–176] This structural complexity allows for the efficient capture and confinement of light, leading to

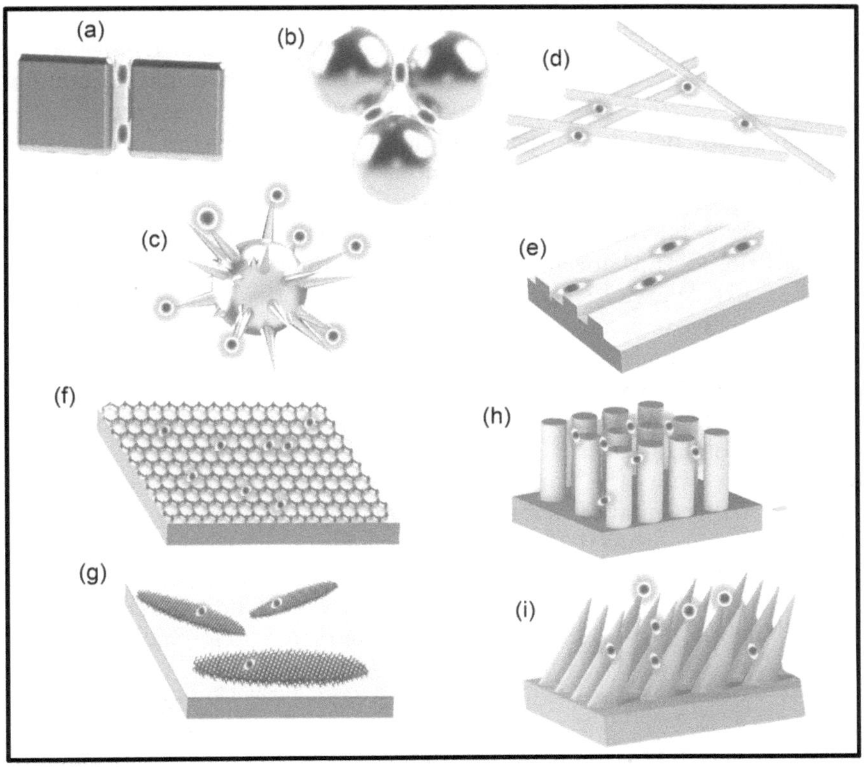

Figure 1.5 Schematic representation of various SERS substrate configurations. One-dimensional structures include (a) nanocube dimer, (b) nanoparticle trimer, and (c) spiked nanosphere. One-dimensional linear structures are shown in (d) nanowires and (e) nanogaps. Two-dimensional substrates are represented by (f) graphene and (g) MXene. Finally, three-dimensional architectures are depicted in (h) nanopillar array and (i) nanocone array. (Reproduced with permission from Ref. [147]. Copyright © 2022 The Authors. American Chemical Society. This publication is licensed under CC-BY-NC-ND 4.0)

enhanced Raman signal generation.[176] As shown by Malerba and Zhang et al., 3D SERS substrates offer enhanced signal amplification and improved spatial resolution compared to their 2D counterparts.[177,178] The 3D architecture of these substrates allows for more effective interaction with light within a laser confocal region, enhancing the Raman signal of target molecules even at extremely low concentrations.[179] Fabrication techniques such as lithography, templating, and physical vapor deposition create 3D SERS substrates.[59,180,181] Challenges in this area include reproducibility, stability, and signal uniformity. Future research focuses on developing multifunctional substrates, integrating them with microfluidic devices for point-of-care diagnostics, and exploring novel materials and fabrication techniques.[182–184]

1.4.5 Hybrid SERS substrates

Hybrid SERS substrates combine the unique properties of different materials to achieve enhanced SERS performance.[185–188] For example, combining metallic nanoparticles with semiconductor nanostructures amplifies the electromagnetic field and provides additional charge transfer mechanisms, leading to improved SERS performance.[153] Gold-silver hybrid nanostructures exhibit enhanced plasmonic properties that boost the Raman signal intensity.[189–191] Furthermore, incorporating 2D materials like graphene enhances the CHEM mechanism due to its unique electronic properties and high surface area.[192] Hybrid nanostructures offer high sensitivity and specificity in biosensing applications.[189] Precise control of nanostructure morphology through fabrication techniques like glancing angle deposition optimizes SERS substrate performance.[58,158] Functionalizing SERS substrates with a surfactant enables selective detection of target analytes. Ligand immobilization and responsive polymers enhance substrate selectivity and versatility.[186,193] Incorporating responsive polymers or molecular gates further enhances substrate selectivity and versatility.[194–196] These modifications are crucial for optimizing the sensitivity and selectivity of SERS-based detection in various applications (Figure 1.6).

In conclusion, the different SERS substrates, including 0D, 1D, 2D, 3D, and hybrid substrates, offer unique characteristics, challenges, and

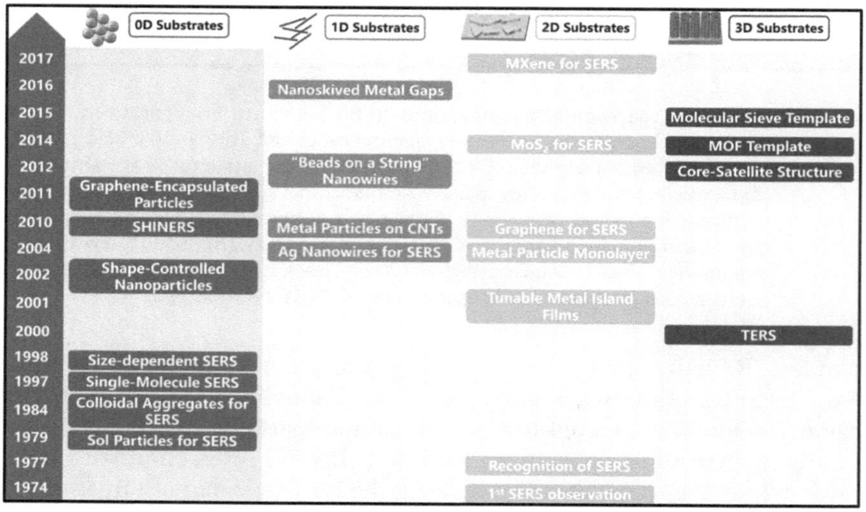

Figure 1.6 Evolution of SERS substrates over the past half-century. This figure highlights key milestones in developing SERS substrates, encompassing 0D nanoparticles, 1D nanowires, 2D metallic and nonmetallic thin films, and 3D nanostructure arrays. (Reproduced with permission from Ref. [147].

potential applications in enhancing the Raman signal for various purposes. Ongoing studies are focused on tackling the issues related to reproducibility, stability, and uniformity of signals. Additionally, they are looking into new materials and methods for manufacturing to improve the efficiency of SERS substrates.

1.5 EMERGING TRENDS AND FUTURE DIRECTIONS

SERS has seen remarkable progress, and it is essential to explore new trends to fully realize its potential in analytical science. Numerous key areas are influencing the future of SERS, focusing on overcoming its current constraints and improving its effectiveness and capabilities.

1. **Nanomaterial Engineering for Enhanced Performance:** One of the key emerging trends in SERS is nanomaterial engineering for enhanced performance. Researchers are actively exploring advanced synthesis techniques such as seed-mediated growth, chemical vapor deposition, and molecular self-assembly to design and synthesize nanomaterials tailored for specific applications.[197–201] These techniques allow precise control over the size, shape, and composition of nanomaterials, leading to improved signal amplification, stability, and reproducibility.[12,202,203] Surface modification strategies like ligand exchange and molecular grafting further optimize the compatibility of nanomaterials with target analytes.[204–206] Ultimately, these advancements aim to improve the sensitivity, specificity, and reliability of SERS-based detection.
2. **Development of Portable and Miniaturized SERS Systems:** The rising need for on-site analysis in healthcare, environmental monitoring, and food safety requires the development of convenient handheld devices designed for field use and settings with limited resources.[207–210] Advances in microfluidics, optics, and detector technologies are leading to the development of convenient handheld devices that are ideal for use in the field and in settings with limited resources with enhanced sensitivity and rapid response times.[209,211,212] These portable SERS systems enable real-time, label-free detection of target analytes in complex samples, facilitating informed decision-making in diverse applications.
3. **Multiplexed and High-Throughput SERS Analysis:** The trend towards combined and efficient SERS analysis caters to the requirement of detecting numerous analytes at the same time in intricate samples.[213] Techniques involving encoded SERS signals, spatially resolved imaging, and spectral analysis enable multiplexed detection with high sensitivity and specificity.[214–216] Various approaches, such as multiplexed nanoparticle probes, barcode-based tagging, and microarray platforms, are

being actively explored to expand the capabilities of SERS for parallel analysis of biomolecules, environmental contaminants, and chemical species.[217–220] This trend holds immense promise for applications like biomarker discovery, drug screening, and molecular diagnostics, where simultaneous detection of multiple targets is crucial.

4. **Smart SERS Substrate Design for Dynamic Sensing Applications:** Smart SERS substrate design is another emerging trend in SERS. By incorporating responsive materials or stimuli-responsive elements, researchers aim to introduce dynamic control over analyte detection and overall sensing performance.[185,221] Responsive polymers, molecular switches, and stimuli-responsive coatings are being utilized to modulate SERS signal intensity, selectivity, and stability in real time.[222–226] These novel substrates find applications in controlled drug release, environmental monitoring, and adaptive sensing platforms, offering dynamic response and tunable sensitivity for optimal performance.[227–232]
5. **Integration of Artificial Intelligence (AI) and ML:** The integration of AI and ML techniques marks a significant shift in SERS research.[233–235] These advancements empower automated data analysis, pattern recognition, and classification of complex SERS spectra.[23,236–240] AI and ML algorithms, such as deep learning, support vector machines, and principal component analysis, accelerate data processing, enhance analytical accuracy, and unlock new insights in biomedical diagnostics, environmental monitoring, and materials characterization.

The rising developments in SERS encompass nanomaterial engineering, portable instrumentation, multiplexed analysis, smart substrate design, and the incorporation of AI. Together, they tackle existing constraints and open the path for creating advanced SERS platforms with improved performance and functionality for applications in healthcare, environmental monitoring, and other fields.

1.6 OUTLOOK AND FUTURE DIRECTIONS

Despite the remarkable advancements in SERS, several key challenges remain that need to be addressed to broaden its application. In this section, we will explore some crucial areas for future research efforts.

1. **Enhancing Reproducibility and Standardization:** One of the major challenges in SERS measurements is the lack of consistency and reliability. To overcome this challenge, it is crucial to establish standardized protocols for SERS measurements, substrate fabrication methods, and data analysis techniques. By doing so, we can ensure greater reproducibility and comparability of SERS results across different laboratories and applications.

2. **Pushing the Limits of Sensitivity:** Enhanced sensitivity in SERS plays a crucial role in identifying minute analytes and early disease indicators. In order to accomplish this goal, researchers need to delve into methods to enhance the spread of hotspots in nanomaterials, examine new material pairings, and integrate cutting-edge signal boosting techniques. These strategies offer potential in extending the boundaries of sensitivity in SERS.
3. **Demystifying Multiplexing and High-Throughput Analysis:** For SERS to detect various analytes at the same time (multiplexing) and analyze extensive sample sets (high-throughput analysis), integrating microfluidics into SERS platforms and creating sophisticated multiplexing techniques are essential. Furthermore, it is important to develop efficient algorithms for processing the collected data for thorough analysis.
4. **Real-Time Monitoring and In Vivo Applications:** Translating SERS into real-time monitoring and in vivo applications poses notable obstacles. Addressing biocompatibility issues, reducing phototoxicity, and creating implantable or wearable SERS sensors are essential for realizing the complete capabilities of SERS in dynamic biological systems. Additional studies are necessary to tackle these hurdles and facilitate the secure and feasible utilization of SERS in real-time monitoring and in vivo studies.
5. **Integration with Complementary Techniques:** By integrating SERS with additional analytical methods like mass spectrometry or electrochemical sensing, a more thorough understanding of analytes can be obtained. The creation of combined platforms that capitalize on the advantages of various techniques shows great potential for enhancing the capabilities and uses of SERS.

Addressing the challenges in reproducibility and standardization, pushing the boundaries of sensitivity, clarifying the complexities of multiplexing and high-throughput analysis, facilitating real-time monitoring and in vivo applications, and integrating SERS with complementary techniques are essential areas for future research endeavors in SERS. By concentrating on these areas, we can overcome current limitations and expand the utilization of SERS in various fields, leading to new developments in biosensing, environmental monitoring, and chemical analysis.

1.7 CONCLUSION

In conclusion, while SERS has already achieved notable advancements in the realm of analytical science, there exist numerous obstacles that must be effectively tackled to facilitate its broader utilization across diverse domains such as biosensing, environmental surveillance, and chemical analysis. The imperative task of surmounting challenges pertaining to reproducibility, enhancement of sensitivity, enabling multiplexing capabilities, real-time monitoring, and seamless integration with complementary methodologies cannot be overstated. The development of multifunctional SERS substrates, which integrate cutting-edge nanomaterials and sophisticated data analysis methodologies, exhibits great potential in addressing these hurdles and ushering in novel pathways for advancement in healthcare, environmental monitoring, and other related areas. The sustained collaborative endeavors and exploration of emerging trends are indispensable in shaping the trajectory of SERS and its influence on various domains.

REFERENCES

1. Pérez-Jiménez, A. I., Lyu, D., Lu, Z., Liu, G. & Ren, B. Surface-enhanced Raman spectroscopy: Benefits, trade-offs and future developments. *Chem. Sci.* **11**, 4563–4577 (2020).
2. Han, X. X., Rodriguez, R. S., Haynes, C. L., Ozaki, Y. & Zhao, B. Surface-enhanced Raman spectroscopy. *Nat. Rev. Methods Prim.* **1**, 1–17 (2022).
3. Eberhardt, K., Stiebing, C., Matthäus, C., Schmitt, M. & Popp, J. Advantages and limitations of Raman spectroscopy for molecular diagnostics: An update. *Expert Rev. Mol. Diag.* **15**, 773–787 (2015).
4. Pathak, C. S. & Kumar, S. *Recent Developments in Atomic Force Microscopy and Raman Spectroscopy for Materials Characterization.* London: Intech Open Limited (2022).
5. Kneipp, K. et al. Single molecule detection using surface-enhanced Raman scattering (SERS). *Phys. Rev. Lett.* **78**, 1667–1670 (1997).
6. Kumar, S., Kumar, P. & Das, A. Surface-enhanced Raman scattering: Introduction and applications. In: M. Kahrizi & P. A. Sohi (eds.), *Recent Advances in Nanophotonics-Fundamentals and Applications* (pp. 137–160). London: IntechOpen Limited (2020). DOI: 10.5772/intechopen.92614
7. Le Ru, E. C., Blackie, E., Meyer, M. & Etchegoin, P. G. Surface enhanced Raman scattering enhancement factors: A comprehensive study. *J. Phys. Chem. C* **111**, 13794–13803 (2007).
8. Kumar, S., Tokunaga, K., Namura, K., Fukuoka, T. & Suzuki, M. Experimental evidence of a twofold electromagnetic enhancement mechanism of surface-enhanced Raman scattering. *J. Phys. Chem. C Nanomater. Interfaces* **124**, 21215–21222 (2020).
9. Guselnikova, O. et al. New trends in nanoarchitectured SERS substrates: Nanospaces, 2D materials, and organic heterostructures. *Small* **18**, e2107182 (2022).
10. Kong, S. M. et al. One-pot platform for the collection and detection of nanoparticles: Flexible surface-enhanced Raman scattering (SERS) substrates with nano-pore structure. *Chem. Eng. J.* **471**, 144753 (2023).
11. Kumar, S., Doi, Y., Namura, K. & Suzuki, M. Plasmonic nanoslit arrays fabricated by serial bideposition: Optical and surface-enhanced Raman scattering study. *ACS Appl. Bio Mater.* **3**, 3226–3235 (2020).

12. Ong, T. T. X., Blanch, E. W. & Jones, O. A. H. Surface Enhanced Raman Spectroscopy in environmental analysis, monitoring and assessment. *Sci. Total Environ.* **720**, 137601 (2020).
13. Chen, M. et al. 2D materials: Excellent substrates for surface-enhanced Raman scattering (SERS) in chemical sensing and biosensing. *Trends Analyt. Chem.* **130**, 115983 (2020).
14. Tahir, M. A., Dina, N. E., Cheng, H., Valev, V. K. & Zhang, L. Surface-enhanced Raman spectroscopy for bioanalysis and diagnosis. *Nanoscale* **13**, 11593–11634 (2021).
15. Koh, E. H. et al. A wearable surface-enhanced Raman scattering sensor for label-free molecular detection. *ACS Appl. Mater. Interfaces* **13**, 3024–3032 (2021).
16. Wilson, N. G., Raveendran, J. & Docoslis, A. Portable identification of fentanyl analogues in drugs using surface-enhanced Raman scattering. *Sens. Actuat. B Chem.* **330**, 129303 (2021).
17. Verma, A. K. & Soni, R. K. Ultrasensitive surface-enhanced Raman spectroscopy detection of explosive molecules with multibranched silver nanostructures. *J. Raman Spectrosc.* **53**, 694–708 (2022).
18. Liao, X. et al. Proximity hybridization-triggered DNA assembly for label-free surface-enhanced Raman spectroscopic bioanalysis. *Anal. Chim. Acta* 1139, 42–49 (2020).
19. Shvalya, V., Filipič, G., Zavašnik, J., Abdulhalim, I. & Cvelbar, U. Surface-enhanced Raman spectroscopy for chemical and biological sensing using nanoplasmonics: The relevance of interparticle spacing and surface morphology. *Appl. Phys. Rev.* **7**, 031307 (2020).
20. Tapio, K. et al. A versatile DNA origami-based plasmonic nanoantenna for label-free single-molecule surface-enhanced Raman spectroscopy. *ACS Nano* **15**, 7065–7077 (2021).
21. Lu, Y., Lin, L. & Ye, J. Human metabolite detection by surface-enhanced Raman spectroscopy. *Mater. Today Bio* **13**, 100205 (2022).
22. Abalde-Cela, S., Wu, L., Teixeira, A., Oliveira, K. & Diéguez, L. Multiplexing liquid biopsy with surface-enhanced Raman scattering spectroscopy. *Adv. Opt. Mater.* **9**, 2001171 (2021).
23. Leong, Y. X. et al. Surface-enhanced Raman scattering (SERS) taster: A machine-learning-driven multireceptor platform for multiplex profiling of wine flavors. *Nano Lett.* **21**, 2642–2649 (2021).
24. Blanco-Formoso, M., Pazos-Perez, N. & Alvarez-Puebla, R. A. Fabrication and SERS properties of complex and organized nanoparticle plasmonic clusters stable in solution. *Nanoscale* **12**, 14948–14956 (2020).
25. Zhang, C. et al. Highly sensitive and reproducible SERS substrates based on ordered micropyramid array and silver nanoparticles. *ACS Appl. Mater. Interfaces* **13**, 29222–29229 (2021).
26. Lee, S. & Choi, I. Fabrication strategies of 3D plasmonic structures for SERS. *Biochip J.* **13**, 30–42 (2019).
27. Chapus, L. et al. Tunable SERS platforms from small nanoparticle 3D superlattices: A comparison between gold, silver, and copper. *Chemphyschem* **18**, 3066–3075 (2017).

28. Liu, H., Yang, L. & Liu, J. Three-dimensional SERS hot spots for chemical sensing: Towards developing a practical analyzer. *Trends Analyt. Chem.* **80**, 364–372 (2016).
29. Betz, J. F., Yu, W. W., Cheng, Y., White, I. M. & Rubloff, G. W. Simple SERS substrates: Powerful, portable, and full of potential. *Phys. Chem. Chem. Phys.* **16**, 2224–2239 (2014).
30. Ouyang, L., Ren, W., Zhu, L. & Irudayaraj, J. Prosperity to challenges: Recent approaches in SERS substrate fabrication. *Rev. Anal. Chem.* **36** (2017).
31. Li, C. et al. Towards practical and sustainable SERS: A review of recent developments in the construction of multifunctional enhancing substrates. *J. Mater. Chem.* **9**, 11517–11552 (2021).
32. Yang, L., Li, P. & Liu, J. Progress in multifunctional surface-enhanced Raman scattering substrate for detection. *RSC Adv.* **4**, 49635–49646 (2014).
33. Huang, J.-A., Zhang, Y.-L., Ding, H. & Sun, H.-B. SERS-enabled lab-on-a-chip systems. *Adv. Opt. Mater.* **3**, 618–633 (2015).
34. Yang, B. et al. Recent development of SERS technology: Semiconductor-based study. *ACS Omega* **4**, 20101–20108 (2019).
35. Tang, X. et al. Alloy engineering allows on-demand design of ultrasensitive monolayer semiconductor SERS substrates. *Nano Lett.* **23**, 7037–7045 (2023).
36. Sharma, B. et al. High-performance SERS substrates: Advances and challenges. *MRS Bull.* **38**, 615–624 (2013).
37. Francis, M. K. et al. Ag nanowires based SERS substrates with very high enhancement factor. *Physica E* **137**, 115080 (2022).
38. Thareja, V., Esfandyarpour, M., Kik, P. G. & Brongersma, M. L. Anisotropic metasurfaces as tunable SERS substrates for 2D materials. *ACS Photonics* **6**, 1996–2004 (2019).
39. Joshi, P. & Santhanam, V. Paper-based SERS active substrates on demand. *RSC Adv.* **6**, 68545–68552 (2016).
40. Marques, P. A. A. P., Nogueira, H. I. S., Pinto, R. J. B., Neto, C. P. & Trindade, T. Silver-bacterial cellulosic sponges as active SERS substrates. *J. Raman Spectrosc.* **39**, 439–443 (2008).
41. Hwang, H. et al. In situ dynamic measurements of the enhanced SERS signal using an optoelectrofluidic SERS platform. *Lab Chip* **11**, 2518–2525 (2011).
42. Qian, K., Yang, L., Li, Z. & Liu, J. A new-type dynamic SERS method for ultrasensitive detection. *J. Raman Spectrosc.* **44**, 21–28 (2013).
43. Kumar, S., Fukuoka, T., Takahashi, R. & Yoshida, M. Highly stable and reproducible Au nanorod arrays for near-infrared optofluidic SERS sensor. *Mater. Lett.* **286**, 129106 (2021).
44. Kumar, S., Lodhi, D. K. & Singh, J. Highly sensitive multifunctional recyclable Ag–TiO_2 nanorod SERS substrates for photocatalytic degradation and detection of dye molecules. *RSC Adv.* **6**, 45120–45126 (2016).
45. Jiang, L. et al. Electrochemically renewable SERS sensor: A new platform for the detection of metabolites involved in peroxide production. *Biosens. Bioelectron.* **175**, 112918 (2021).
46. Laing, S., Gracie, K. & Faulds, K. Multiplex in vitro detection using SERS. *Chem. Soc. Rev.* **45**, 1901–1918 (2016).

47. Zhou, L. et al. Size-tunable gold aerogels: A durable and misfocus-tolerant 3D substrate for multiplex SERS detection. *Adv. Opt. Mater.* **9**, 2100352 (2021).
48. Iakab, S.-A. et al. SALDI-MS and SERS multimodal imaging: One nanostructured substrate to rule them both. *Anal. Chem.* **94**, 2785–2793 (2022).
49. Liu, J., Siavash Moakhar, R., Mahshid, S., Vasefi, F. & Wachsmann-Hogiu, S. Multimodal electrochemical and SERS platform for chlorfenapyr detection. *Appl. Surf. Sci.* **566**, 150617 (2021).
50. Zheng, Y. et al. Phase controlled SERS enhancement. *Sci. Rep.* **9**, 744 (2019).
51. Chou, J.-A. et al. Organic electrochemical transistors/SERS-active hybrid biosensors featuring gold nanoparticles immobilized on thiol-functionalized PEDOT films. *Front. Chem.* **7**, 281 (2019).
52. Kumar, S., Namura, K. & Suzuki, M. Proposal for a gel-based SERS sensor. *Proc. SPIE*, **10894**, 1089414 (2019).
53. Kumar, S., Taneichi, T., Fukuoka, T., Namura, K. & Suzuki, M. Study on transport of molecules in gel by surface-enhanced Raman spectroscopy. *Cellulose* **28**, 10803 (2021).
54. Yang, W. et al. Recyclable Ag-deposited TiO2 SERS substrate for ultrasensitive malachite green detection. *ACS Omega* **6**, 27271–27278 (2021).
55. Trojanowicz, M. & Szewczyńska, M. Biosensing in high-performance chemical separations. *Trends Analyt. Chem.* **24**, 92–106 (2005).
56. Nie, S. & Emory, S. R. Probing single molecules and single nanoparticles by surface-enhanced Raman scattering. *Science* **275**, 1102–1106 (1997).
57. Hoang, L. T., Van Pham, H. & Nguyen, M. T. T. Investigation of the factors influencing the surface-enhanced Raman scattering activity of silver nanoparticles. *J. Electron. Mater.* **49**, 1864–1871 (2020).
58. Kumar, S., Gahlaut, S. K. & Singh, J. P. Sculptured thin films: Overcoming the limitations of surface-enhanced Raman scattering substrates. *Appl. Surf. Sci. Adv.* **12**, 100322 (2022).
59. Yadav, S., Senapati, S., Kumar, S., Gahlaut, S. K. & Singh, J. P. GLAD based advanced nanostructures for diversified biosensing applications: Recent progress. *Biosensors* **12** (2022).
60. Kumar, S. & Seo, S. Plasmonic sensors: A new frontier in nanotechnology. *Biosensors (Basel)* **13**, 385 (2023).
61. Haynes, C. L. & Van Duyne, R. P. Nanosphere lithography: A versatile nanofabrication tool for studies of size-dependent nanoparticle optics. *J. Phys. Chem. B* **105**, 5599–5611 (2001).
62. Coronado, E. A., Encina, E. R. & Stefani, F. D. Optical properties of metallic nanoparticles: Manipulating light, heat and forces at the nanoscale. *Nanoscale* **3**, 4042–4059 (2011).
63. Katyal, J. Comparative study between different plasmonic materials and nanostructures for sensor and SERS application. In: C. D. Geddes (ed.), *Reviews in Plasmonics 2017,* Springer International Publishing, Cham, pp. 77–108 (2019). doi:10.1007/978-3-030-18834-4_4.
64. Xie, L. et al. State of the art in flexible SERS sensors toward label-free and onsite detection: From design to applications. *Nano Res.* **15**, 4374–4394 (2022).

65. Demydov, P. V., Lopatynskyi, A. M., Hudzenko, I. I. & Chegel, V. I. The approaches for localized surface plasmon resonance wavelength position tuning: Short review. *Semicond. Phys. Quantum Electron. Optoelectron.* **24**, 304–311 (2021).
66. Dieringer, J. A. et al. Surface enhanced Raman spectroscopy: New materials, concepts, characterization tools, and applications. *Faraday Discuss.* **132**, 9–26 (2006).
67. Bochenkov, V. et al. Applications of plasmonics: General discussion. *Faraday Discuss.* **178**, 435–466 (2015).
68. Etchegoin, P. G., & Le Ru, E. C. A perspective on single molecule SERS: current status and future challenges. *Phys. Chem. Chem. Phys.* **10**, 6079–6089 (2008).
69. Fu, J. & Zhao, Y. Au nanoparticle based localized surface plasmon resonance substrates fabricated by dynamic shadowing growth. *Nanotechnology* **21**, 175303 (2010).
70. Willets, K. A. & Van Duyne, R. P. Localized surface plasmon resonance spectroscopy and sensing. *Annu. Rev. Phys. Chem.* **58**, 267–297 (2007).
71. Tharion, J., Satija, J. & Mukherji, S. Facile synthesis of size-tunable silver nanoparticles by heteroepitaxial growth method for efficient NIR SERS. *Plasmonics* **10**, 753–763 (2015).
72. Boyack, R. & Le Ru, E. C. Investigation of particle shape and size effects in SERS using T-matrix calculations. *Phys. Chem. Chem. Phys.* **11**, 7398–7405 (2009).
73. Benz, F. et al. SERS of individual nanoparticles on a mirror: Size does matter, but so does shape. *J. Phys. Chem. Lett.* **7**, 2264–2269 (2016).
74. Tiwari, V. S. et al. Non-resonance SERS effects of silver colloids with different shapes. *Chem. Phys. Lett.* **446**, 77–82 (2007).
75. Vigderman, L. & Zubarev, E. R. Starfruit-shaped gold nanorods and nanowires: Synthesis and SERS characterization. *Langmuir* **28**, 9034–9040 (2012).
76. Lv, Y. et al. Surface morphology evolution behavior and SERS performance of Mo-Ag-Cu-Co films. *Appl. Surf. Sci.* **604** (2022).
77. Oh, Y.-J., Kang, M., Park, M. & Jeong, K.-H. Engineering hot spots on plasmonic nanopillar arrays for SERS: A review. *Biochip J.* **10**, 297–309 (2016).
78. Brandl, D. W., Mirin, N. A. & Nordlander, P. Plasmon modes of nanosphere trimers and quadrumers. *J. Phys. Chem. B* **110**, 12302–12310 (2006).
79. Zhang, Z.-Y. & Zhao, Y.-P. Tuning the optical absorption properties of Ag nanorods by their topologic shapes: A discrete dipole approximation calculation. *Appl. Phys. Lett.* **89**, 023110 (2006).
80. Jiao, J. et al. Polarization-dependent SERS at differently oriented single gold nanorods. *Chemphyschem* **13**, 952–958 (2012).
81. Pastoriza-Santos, I. & Liz-Marzán, L. M. Colloidal silver nanoplates: State of the art and future challenges. *J. Mater. Chem.* **18**, 1724–1737 (2008).
82. Keast, V. J. et al. Higher order plasmonic modes excited in Ag triangular nanoplates by an electron beam. *Plasmonics* **11**, 1081–1086 (2016).
83. Mock, J. J., Barbic, M., Smith, D. R., Schultz, D. A. & Schultz, S. Shape effects in plasmon resonance of individual colloidal silver nanoparticles. *J. Chem. Phys.* **116**, 6755–6759 (2002).

84. Keating, M. et al. Ordered silver and copper nanorod arrays for enhanced Raman scattering created via guided oblique angle deposition on polymer. *J. Phys. Chem. C* **118**, 4878–4884 (2014).
85. Qian, C. et al. Highly-ordered, 3D petal-like array for surface-enhanced Raman scattering. *Small* 7, 1800–1806 (2011).
86. Wu, X. et al. Improved SERS nanoparticles for direct detection of circulating tumor cells in the blood. *ACS Appl. Mater. Interfaces* 7, 9965–9971 (2015).
87. Schmidt, T. M., Frederiksen, M., Bochenkov, V. & Sutherland, D. S. Exploring plasmonic coupling in hole-cap arrays. *Beilstein J. Nanotechnol.* **6**, 1–10 (2015).
88. Muri, H. I. & Hjelme, D. R. LSPR coupling and distribution of interparticle distances between nanoparticles in hydrogel on optical fiber end face. *Sensors* **17**, 2723 (2017).
89. Kim, S.-M., Zhang, W. & Cunningham, B. T. Coupling discrete metal nanoparticles to photonic crystal surface resonant modes and application to Raman spectroscopy. *Opt. Express* **18**, 4300 (2010).
90. Aizpurua, J. & Hillenbrand, R. Localized surface plasmons: Basics and applications in field-enhanced spectroscopy. In: *Plasmonics,* pp. 151–176 (2012). doi:10.1007/978-3-642-28079-5_5.
91. Hao, F. et al. Symmetry breaking in plasmonic nanocavities: Subradiant LSPR sensing and a tunable Fano resonance. *Nano Lett.* **8**, 3983–3988 (2008).
92. Link, S. & El-Sayed, M. A. Spectral properties and relaxation dynamics of surface plasmon electronic oscillations in gold and silver nanodots and nanorods. *J. Phys. Chem. B* **103**, 8410–8426 (1999).
93. Krenn, J. R. et al. Design of multipolar plasmon excitations in silver nanoparticles. *Appl. Phys. Lett.* 77, 3379–3381 (2000).
94. Sönnichsen, C. et al. Drastic reduction of plasmon damping in gold nanorods. *Phys. Rev. Lett.* **88** (2002).
95. Hanarp, P., Käll, M. & Sutherland, D. S. Optical properties of short range ordered arrays of nanometer gold disks prepared by colloidal lithography. *J. Phys. Chem. B* **107**, 5768–5772 (2003).
96. Oldenburg, S. J., Westcott, S. L., Averitt, R. D. & Halas, N. J. Surface enhanced Raman scattering in the near infrared using metal nanoshell substrates. *J. Chem. Phys.* **111**, 4729–4735 (1999).
97. Aizpurua, J. et al. Optical properties of gold nanorings. *Phys. Rev. Lett.* **90**, 057401 (2003).
98. Wang, H., Brandl, D. W., Le, F., Nordlander, P. & Halas, N. J. Nanorice: A hybrid plasmonic nanostructure. *Nano Lett.* **6**, 827–832 (2006).
99. Sherry, L. J. et al. Localized surface plasmon resonance spectroscopy of single silver nanocubes. *Nano Lett.* **5**, 2034–2038 (2005).
100. Nehl, C. L., Liao, H. & Hafner, J. H. Optical properties of star-shaped gold nanoparticles. *Nano Lett.* **6**, 683–688 (2006).
101. Knight, M. W. & Halas, N. J. Nanoshells to nanoeggs to nanocups: Optical properties of reduced symmetry core–shell nanoparticles beyond the quasistatic limit. *New J. Phys.* **10** (2008).
102. Mirin, N. A. & Halas, N. J. Light-bending nanoparticles. *Nano Lett.* **9**, 1255–1259 (2009).

103. Ziegler, J. I. & Haglund, R. F. Plasmonic response of nanoscale spirals. *Nano Lett.* **10**, 3013–3018 (2010).
104. Bukasov, R. & Shumaker-Parry, J. S. Highly tunable infrared extinction properties of gold nanocrescents. *Nano Lett.* 7, 1113–1118 (2007).
105. Jin, R. et al. Photoinduced conversion of silver nanospheres to nanoprisms. *Science* **294**, 1901–1903 (2001).
106. Myroshnychenko, V. et al. Modelling the optical response of gold nanoparticles. *Chem. Soc. Rev.* **37** (2008).
107. Rindzevicius, T. et al. Nanohole plasmons in optically thin gold films. *J. Phys. Chem. C Nanomater. Interfaces* **111**, 1207–1212 (2006).
108. Rigneault, H. et al. Enhancement of single-molecule fluorescence detection in subwavelength apertures. *Phys. Rev. Lett.* **95** (2005).
109. Sukharev, M., Sung, J., Spears, K. G. & Seideman, T. Optical properties of metal nanoparticles with no center of inversion symmetry: Observation of volume plasmons. *Phys. Rev. B Condens. Matter Mater. Phys.* **76** (2007).
110. Kumar, S., Goel, P., Singh, D. P. & Singh, J. P. Highly sensitive superhydrophobic Ag nanorods array substrates for surface enhanced fluorescence studies. *Appl. Phys. Lett.* (2014).
111. Guerra Hernández, L. A., Reynoso, A. A. & Fainstein, A. Does the chemical contribution have a secondary role in SERS? *J. Opt. Soc. Am. B* **40** (2023).
112. Kim, C., Hong, B. & Choi, W. Surface-enhanced Raman spectroscopy (SERS) investigation of a 3D plasmonic architecture utilizing Ag nanoparticles-embedded functionalized carbon nanowall. *Nanomaterials (Basel)* **13**, 2617 (2023).
113. Xu, Y., Zhang, Y., Li, C., Ye, Z. & Bell, S. E. J. SERS as a probe of surface chemistry enabled by surface-accessible plasmonic nanomaterials. *Acc. Chem. Res.* **56**, 2072–2083 (2023).
114. Vilkevičius, K., Ignatjev, I., Selskis, A., Niaura, G. & Stankevičius, E. Tuning SERS performance through the laser-induced morphology changes of gold nanostructures. *Appl. Surf. Sci.* **660**, 160003 (2024).
115. Kukushkin, V. I., Astrakhantseva, A. S. & Morozova, E. N. Influence of the morphology of metal nanoparticles deposited on surfaces of silicon oxide on the optical properties of SERS Substrates. *Bull. Russ. Acad. Sci. Phys.* **85**, 133–140 (2021).
116. Feng, Z., Jia, Y. & Cui, H. Engineering the surface roughness of the gold nanoparticles for the modulation of LSPR and SERS. *J. Colloid Interface Sci.* **672**, 1–11 (2024).
117. Ding, S. et al. Surface-roughness-adjustable Au nanorods with strong plasmon absorption and abundant hotspots for improved SERS and photothermal performances. *Nano Res.* **15**, 2715–2721 (2021).
118. Pham, P., Goel, P., Kumar, S. & Yadav, K. (eds.). *21st Century Surface Science: A Handbook*. London: IntechOpen Limited (2020).
119. Jena, T. et al. Evidence for intrinsic defects and nanopores as hotspots in 2D PdSe2 dendrites for plasmon-free SERS substrate with a high enhancement factor. *Npj 2D Mater. Appl.* 7, 8 (2023).
120. Sun, H., Wang, X., Wu, P., Jiang, H. & Tang, J. Nonstoichiometric tungsten oxide nanosheets with abundant oxygen vacancies for defects-driven SERS sensing. *J. Raman Spectrosc.* **53**, 1880–1889 (2022).

121. Ye, M. & Tang, S. Construction of SERS chip based on silver nanoparticles and detection of sports doping β-agonists. *Alex. Eng. J.* **83**, 134–139 (2023).
122. Dharmalingam, P., Venkatakrishnan, K. & Tan, B. An atomic-defect enhanced Raman scattering (DERS) quantum probe for molecular level detection – Breaking the SERS barrier. *Appl. Mater. Today* **16**, 28–41 (2019).
123. Yang, C.-W., Zhang, X., Yuan, L., Wang, Y.-K. & Sheng, G.-P. Deciphering the microheterogeneous repartition effect of environmental matrix on surface-enhanced Raman spectroscopy (SERS) analysis for pollutants in natural waters. *Water Res.* **232**, 119668 (2023).
124. Becca, J. C., Chen, X. & Jensen, L. A discrete interaction model/quantum mechanical method for simulating surface-enhanced Raman spectroscopy in solution. *J. Chem. Phys.* **154**, 224705 (2021).
125. Chen, R. & Jensen, L. Interpreting chemical enhancements of surface-enhanced Raman scattering. *Chem. Phys. Rev.* **4**, 021305 (2023).
126. Song, C. X. et al. Surface-enhanced Raman spectroscopy substrate time stability improvement using an external oxygen barrier method. *Appl. Spectrosc.* **78**, 289–295 (2024).
127. Wang, T. et al. Dynamic modulation of a surface-enhanced Raman scattering signal by a varying magnetic field. *Opt. Express* **31** (2023).
128. Ding, S.-Y., You, E.-M., Tian, Z.-Q. & Moskovits, M. Electromagnetic theories of surface-enhanced Raman spectroscopy. *Chem. Soc. Rev.* **46**, 4042–4076 (2017).
129. Cong, S., Liu, X., Jiang, Y., Zhang, W. & Zhao, Z. Surface enhanced Raman scattering revealed by interfacial charge-transfer transitions. *Innovation (NY)* **1**, 100051 (2020).
130. Itoh, T. et al. Toward a new era of SERS and TERS at the nanometer scale: From fundamentals to innovative applications. *Chem. Rev.* **123**, 1552–1634 (2023).
131. Grasseschi, D. & Toma, H. E. The SERS effect in coordination chemistry. *Coord. Chem. Rev.* **333**, 108–131 (2017).
132. Moskovits, M. Surface-enhanced Raman spectroscopy: A brief retrospective. *J. Raman Spectrosc.* **36**, 485–496 (2005).
133. Meyer, M., Le Ru, E. C. & Etchegoin, P. G. Self-limiting aggregation leads to long-lived metastable clusters in colloidal solutions. *J. Phys. Chem. B* **110**, 6040–6047 (2006).
134. Shanmukh, S. et al. Identification and classification of respiratory syncytial virus (RSV) strains by surface-enhanced Raman spectroscopy and multivariate statistical techniques. *Anal. Bioanal. Chem.* **390**, 1551–1555 (2008).
135. Driskell, J. D. et al. Rapid microRNA (miRNA) detection and classification via surface-enhanced Raman spectroscopy (SERS). *Biosens. Bioelectron.* **24**, 917–922 (2008).
136. Vo-Dinh, T. *Biomedical Photonics Handbook*. (2003). doi:10.1201/9780203008997.
137. Xu, H., Bjerneld, E. J., Käll, M. & Börjesson, L. Spectroscopy of single hemoglobin molecules by surface enhanced Raman scattering. *Phys. Rev. Lett.* **83**, 4357–4360 (1999).
138. Zhang, X., Chen, Y. L., Liu, R.-S. & Tsai, D. P. Plasmonic photocatalysis. *Rep. Prog. Phys.* **76**, 046401 (2013).

139. Plou, J. et al. Multiplex SERS detection of metabolic alterations in tumor extracellular media. *Adv. Funct. Mater.* **30**, 1910335 (2020).
140. Zhang, W. et al. Multiplex SERS-based lateral flow immunosensor for the detection of major mycotoxins in maize utilizing dual Raman labels and triple test lines. *J. Hazard. Mater.* **393**, 122348 (2020).
141. Yuan, Y. et al. Optical trapping-assisted SERS platform for chemical and biosensing applications: Design perspectives. *Coord. Chem. Rev.* **339**, 138–152 (2017).
142. Dai, X. et al. Optical tweezers-controlled hotspot for sensitive and reproducible surface-enhanced Raman spectroscopy characterization of native protein structures. *Nat. Commun.* **12**, 1292 (2021).
143. Huo, Y., Yang, Z., Wilson, T. & Jiang, C. Recent progress in SERS-based anti-counterfeit labels. *Adv. Mater. Interfaces* **9**, 2200201 (2022).
144. Liu, S. et al. Multi-functional plasmonic fabrics: A flexible SERS substrate and anti-counterfeiting security labels with tunable encoding information. *Appl. Surf. Sci.* **567**, 150861 (2021).
145. Li, D. et al. Anti-counterfeiting SERS security labels derived from silver nanoparticles and aryl diazonium salts. *Nanosc. Adv.* **4**, 5037–5043 (2022).
146. Chen, J. et al. Progress of microfluidics combined with SERS technology in the trace detection of harmful substances. *Chemosensors* **10**, 449 (2022).
147. Long, L., Ju, W., Yang, H.-Y. & Li, Z. Dimensional design for surface-enhanced Raman spectroscopy. *ACS Mater. Au* **2**, 552–575 (2022).
148. Liu, H., Gao, X., Xu, C. & Liu, D. SERS tags for biomedical detection and bioimaging. *Theranostics* **12**, 1870–1903 (2022).
149. Wieszczycka, K. et al. Surface functionalization: The way for advanced applications of smart materials. *Coord. Chem. Rev.* **436**, 213846 (2021).
150. Abraham, S. et al. Carbon nanostructure (0–3 dimensional) supported isolated gold nanoparticles as an effective SERS substrate. *Sens. Actuat. B Chem.* **273**, 455–465 (2018).
151. Wang, Z., Hu, T., Liang, R. & Wei, M. Application of zero-dimensional nanomaterials in biosensing. *Front. Chem.* **8**, 320 (2020).
152. Schütz, M. & Schlücker, S. Molecularly linked 3D plasmonic nanoparticle core/satellite assemblies: SERS nanotags with single-particle Raman sensitivity. *Phys. Chem. Chem. Phys.* **17**, 24356–24360 (2015).
153. Langer, J. et al. Present and future of surface-enhanced Raman scattering. *ACS Nano* **14**, 28–117 (2020).
154. Di Fabrizio, E. et al. Roadmap on biosensing and photonics with advanced nano-optical methods. *J. Opt.* **18**, 1–27.
155. Singh, D. P., Kumar, S. & Singh, J. P. Morphology dependent surface enhanced fluorescence study on silver nanorod arrays fabricated by glancing angle deposition. *RSC Adv.* **5**, 31341–31346 (2015).
156. Driskell, J. D. et al. The use of aligned silver nanorod arrays prepared by oblique angle deposition as surface enhanced Raman scattering substrates. *J. Phys. Chem. C Nanomater. Interfaces* **112**, 895–901 (2008).
157. Ai, B. & Zhao, Y. Glancing angle deposition meets colloidal lithography: A new evolution in the design of nanostructures. *Nanophotonics* **8**, 1–26 (2018).
158. Ming, T. et al. Strong polarization dependence of plasmon-enhanced fluorescence on single gold nanorods. *Nano Lett.* **9**, 3896–3903 (2009).

159. Suzuki, M. & Taga, Y. Anisotropy in the optical absorption of Ag--SiO2 thin films with oblique columnar structures. *J. Appl. Phys.* **71**, 2848–2854 (1992).
160. Sinha, S. S., Jones, S., Pramanik, A. & Ray, P. C. Nanoarchitecture based SERS for biomolecular fingerprinting and label-free disease markers diagnosis. *Acc. Chem. Res.* **2016**, 2725–2735 (2016).
161. Yu, Y. et al. Roadmap for single-molecule surface-enhanced Raman spectroscopy. *AP* **2**, 014002 (2020).
162. Graham, D., Moskovits, M. & Tian, Z.-Q. SERS: Facts, figures and the future. *Chem. Soc. Rev.* **46**, 3864–3865 (2017).
163. West, P. R. et al. Searching for better plasmonic materials. *Laser Photon. Rev.* **4**, 795–808 (2010).
164. Schlücker, S. Surface-enhanced Raman spectroscopy: Concepts and chemical applications. *Angew. Chem. Int. Ed Engl.* **53**, 4756–4795 (2014).
165. Rajput, A., Kumar, S. & Singh, J. Vertically standing nanoporous Al-Ag zig-zag silver nanorod arrays for highly active SERS substrates. *In Anal.* **142**, 3959–3966 (2017).
166. Ye, Z., Li, C., Chen, Q., Xu, Y. & Bell, S. E. J. Self-assembly of colloidal nanoparticles into 2D arrays at water-oil interfaces: Rational construction of stable SERS substrates with accessible enhancing surfaces and tailored plasmonic response. *Nanoscale* **13**, 5937–5953 (2021).
167. Su, S. et al. Creating SERS hot spots on MoS(2) nanosheets with in situ grown gold nanoparticles. *ACS Appl. Mater. Interfaces* **6**, 18735–18741 (2014).
168. Tao, L. et al. Enhancing light-matter interaction in 2D materials by optical micro/nano architectures for high-performance optoelectronic devices. *InfoMat* **3**, 36–60 (2020).
169. Maurer, T. et al. Enhancing LSPR sensitivity of au gratings through graphene coupling to Au film. *Plasmonics* **9**, 507–512 (2014).
170. Vianna, P. G. et al. Graphene oxide/gold nanorod nanocomposite for stable surface-enhanced Raman spectroscopy. *ACS Photonics* **3**, 1027–1035 (2016).
171. Ling, X. et al. Can graphene be used as a substrate for Raman enhancement? *Nano Lett.* **10**, 553–561 (2010).
172. Polyushkin, D. K. et al. Graphene as a substrate for plasmonic nanoparticles. *J. Opt.* **15**, 114001 (2013).
173. Kumar, S. et al. A facile method for fabrication of buckled PDMS silver nanorod arrays as active 3D SERS cages for bacterial sensing. *Chem. Commun.* **51**, 12411–12414 (2015).
174. Yarbakht, M., Nikkhah, M., Moshaii, A., Abbasian, S. & Dellith, A. Fabrication of silver chevron arrays as an efficient and stable SERS substrate: Implications in biological sensing. *Plasmonics* **13**, 715–726 (2018).
175. Yilmaz, M. et al. Combining 3-D plasmonic gold nanorod arrays with colloidal nanoparticles as a versatile concept for reliable, sensitive, and selective molecular detection by SERS. *Phys. Chem. Chem. Phys.* **16**, 5563–5570 (2014).
176. Phyo, J. B. et al. Label-free SERS analysis of urine using a 3D-stacked AgNW-glass fiber filter sensor for the diagnosis of pancreatic cancer and prostate cancer. *Anal. Chem.* **93**, 3778–3785 (2021).
177. Malerba, M. et al. 3D vertical nanostructures for enhanced infrared plasmonics. *Sci. Rep.* **5**, 16436 (2015).

178. Zhang, Q., Lee, Y. H., Phang, I. Y., Lee, C. K. & Ling, X. Y. Hierarchical 3D SERS substrates fabricated by integrating photolithographic microstructures and self-assembly of silver nanoparticles. *Small* **10**, 2703–2711 (2014).
179. Hu, W., Xia, L., Hu, Y. & Li, G. Recent progress on three-dimensional substrates for surface-enhanced Raman spectroscopic analysis. *Microchem. J.* **172**, 106908 (2022).
180. Gao, S., Guo, Z. & Liu, Z. Recent advances in rational design and engineering of signal-amplifying substrates for surface-enhanced Raman scattering-based bioassays. *Chemosensors* **11**, 461 (2023).
181. Zhao, Y. et al. A SERS sensor based on 3D nanocone forests capable of intelligent classification of aquatic product dyes. *J. Mater. Chem. C Mater. Opt. Electron. Devices* **11**, 14237–14247 (2023).
182. Yadav, S. et al. Portable and sensitive Ag nanorods based SERS platform for rapid HIV-1 detection and tropism determination. *Colloids Surf. B Biointerfaces* **198**, 111477 (2021).
183. Xia, L. & Li, G. Recent progress of microfluidics in surface-enhanced Raman spectroscopic analysis. *J. Sep. Sci.* **44**, 1752–1768 (2021).
184. Schierhorn, M., Lee, S. J., Boettcher, S. W., Stucky, G. D. & Moskovits, M. Metal–silica hybrid nanostructures for surface-enhanced Raman spectroscopy. *Adv. Mater.* **18**, 2829–2832 (2006).
185. Hu, B., Pu, H. & Sun, D.-W. Multifunctional cellulose based substrates for SERS smart sensing: Principles, applications and emerging trends for food safety detection. *Trends Food Sci. Technol.* **110**, 304–320 (2021).
186. Huang, Z. et al. Large-area Ag nanorod array substrates for SERS: AAO template-assisted fabrication, functionalization, and application in detection PCBs. *J. Raman Spectrosc.* **44**, 240–246 (2013).
187. Solís, D. M., Taboada, J. M., Obelleiro, F., Liz-Marzán, L. M. & García de Abajo, F. J. Optimization of Nanoparticle-Based SERS Substrates through Large-Scale Realistic Simulations. *ACS Photonics* **4**, 329–337 (2017).
188. Kumar, S., Namura, K., Kumaki, D., Tokito, S. & Suzuki, M. Highly reproducible, large scale inkjet-printed Ag nanoparticles-ink SERS substrate. *Results in Materials* **8**, 100139 (2020).
189. Fateixa, S., Nogueira, H. I. S. & Trindade, T. Hybrid nanostructures for SERS: Materials development and chemical detection. *Phys. Chem. Chem. Phys.* **17**, 21046–21071 (2015).
190. Gambucci, M., Cambiotti, E., Sassi, P. & Latterini, L. Multilayer gold-silver bimetallic nanostructures to enhance SERS detection of drugs. *Molecules* **25**, 3405 (2020).
191. Yin, Y. et al. Enhanced plasmonic resonance characteristics of AgNRs–gold film hybrid system. *Front. Chem.* **8** (2021).
192. Zhou, Q. et al. Graphene-based surface-enhanced Raman scattering (SERS) sensing: Bibliometrics based analysis and review. *Chemosensors (Basel)* **10**, 317 (2022).
193. Chen, L. et al. Ultrasensitive surface-enhanced Raman scattering nanosensor for mercury ion detection based on functionalized silver nanoparticles. *RSC Adv.* **4**, 15055–15060 (2014).
194. Indrasekara, A. S. D. S., Wadams, R. C. & Fabris, L. Ligand exchange on gold nanorods: Going back to the future. *Part. Part. Syst. Charact.* **31**, 819–838 (2014).

195. Pellas, V. et al. Gold Nanorods for LSPR Biosensing: Synthesis, Coating by Silica, and Bioanalytical Applications. *Biosensors (Basel)* **10**, 146 (2020).
196. Kodiyath, R. et al. Assemblies of silver nanocubes for highly sensitive SERS chemical vapor detection. *J. Mater. Chem. A Mater. Energy Sustain.* **1**, 2777 (2013).
197. Vinod, M. & Gopchandran, K. G. Ag@Au core–shell nanoparticles synthesized by pulsed laser ablation in water: Effect of plasmon coupling and their SERS performance. *Spectrochim. Acta A Mol. Biomol. Spectrosc.* **149**, 913–919 (2015).
198. Zhang, M., Zhao, A., Wang, D. & Sun, H. Hierarchically assembled NiCo@SiO2@Ag magnetic core-shell microspheres as highly efficient and recyclable 3D SERS substrates. *Analyst* **140**, 440–448 (2015).
199. Liedtke, S., Grüner, C., Lotnyk, A. & Rauschenbach, B. Glancing angle deposition of sculptured thin metal films at room temperature. *Nanotechnology* **28**, 385604 (2017).
200. Nikoobakht, B. & El-Sayed, M. A. Preparation and growth mechanism of gold nanorods (NRs) using seed-mediated growth method. *Chem. Mater.* **15**, 1957–1962 (2003).
201. Fu, J., Park, B. & Zhao, Y. Nanorod-mediated surface plasmon resonance sensor based on effective medium theory. *Appl. Opt.* **48**, 4637–4649 (2009).
202. Rong, K., Liu, H., Shi, K. & Chen, J. Pattern-assisted stacking colloidal quantum dots for photonic integrated circuits. *Nanoscale* **11**, 13885–13893 (2019).
203. Zhang, J., Joshi, P., Zhou, Y., Ding, R. & Zhang, P. Quantitative SERS-based DNA detection assisted by magnetic microspheres. *Chem. Commun.* **51**, 15284–15286 (2015).
204. Alexander, K. D. et al. A high-throughput method for controlled hot-spot fabrication in SERS-active gold nanoparticle dimer arrays. *J. Raman Spectrosc.* **40**, 2171–2175 (2009).
205. Park, S.-G. et al. Surface energy-controlled SERS substrates for molecular concentration at plasmonic nanogaps. *Adv. Funct. Mater.* **27**, 1703376 (2017).
206. Kalachyova, Y. et al. The effect of silver grating and nanoparticles grafting for LSP–SPP coupling and SERS response intensification. *J. Phys. Chem. C* **120**, 10569–10577 (2016).
207. Shin, S. et al. Field-portable seawater toxicity monitoring platform using lens-free shadow imaging technology. *Water Res.* **230**, 119585 (2023).
208. Seo, D. et al. Field-portable leukocyte classification device based on lens-free shadow imaging technique. *Biosensors* **12**, 47 (2022).
209. Villa, J. E. L. & Poppi, R. J. A portable SERS method for the determination of uric acid using a paper-based substrate and multivariate curve resolution. *Analyst* **141**, 1966–1972 (2016).
210. Kolomijeca, A., Kronfeldt, H. & Kwon, Y.-H. A portable surface enhanced Raman (SERS) sensor system applied for seawater and sediment investigations on an arctic sea-trial. *Int. J. Offshore Polar* **23**, 161–165 (2013).
211. Cabrera, F. C. et al. Portable smart films for ultrasensitive detection and chemical analysis using SERS and SERRS. *J. Raman Spectrosc.* **43**, 474–477 (2012).
212. Yang, H., Shin, S., Seo, D., Park, J. & Seo, S. Detection of particulate matters with a field-portable microscope using side-illuminated total internal reflection. *Sensors* **21** (2021).

213. Zhang, Y. et al. Combining multiplex SERS nanovectors and multivariate analysis for in situ profiling of circulating tumor cell phenotype using a microfluidic chip. *Small* **14**, e1704433 (2018).
214. Cha, M. G. et al. High-throughput multiplex analysis method based on fluorescence–SERS quantum dot-embedded silver bumpy nanoprobes. *Appl. Surf. Sci.* **558**, 149787 (2021).
215. Maiti, K. K. et al. Multiplex targeted in vivo cancer detection using sensitive near-infrared SERS nanotags. *Nano Today* 7, 85–93 (2012).
216. Ngo, H. T., Wang, H.-N., Burke, T., Ginsburg, G. S. & Vo-Dinh, T. Multiplex detection of disease biomarkers using SERS molecular sentinel-on-chip. *Anal. Bioanal. Chem.* **406**, 3335–3344 (2014).
217. Guarrotxena, N. & Bazan, G. C. Antitags: SERS-encoded nanoparticle assemblies that enable single-spot multiplex protein detection. *Adv. Mater.* **26**, 1941–1946 (2014).
218. Oh, Y.-J. & Jeong, K.-H. Optofluidic SERS chip with plasmonic nanoprobes self-aligned along microfluidic channels. *Lab Chip* **14**, 865–868 (2014).
219. Namura, K., Hanai, S., Kondo, S., Kumar, S. & Suzuki, M. Gold micropetals self-assembled by shadow-sphere lithography for optofluidic control (back cover). *Adv. Mater. Interfaces* **9**, 2270104 (2022).
220. Namura, K., Hanai, S., Kumar, S. & Suzuki, M. Plasmonic structures self-assembled by glancing angle deposition for optofluidic devices. In: *35 (SPIE-Intl Soc Optical Eng, 2021)*. doi:10.1117/12.2593711.
221. Hu, F. et al. Smart Liquid SERS Substrates based on Fe3O4/Au Nanoparticles with Reversibly Tunable Enhancement Factor for Practical Quantitative Detection. *Sci. Rep.* **4**, 1–10 (2014).
222. Shen, Y. et al. Dual stimulus-responsive core-satellite SERS nanoprobes for reactive oxygen species sensing during autophagy. *Talanta* **250**, 123712 (2022).
223. Liu, X. et al. Temperature- and pH-tunable plasmonic properties and SERS efficiency of the silver nanoparticles within the dual stimuli-responsive microgels. *J. Mater. Chem.* **2**, 7326–7335 (2014).
224. Filippin, N. et al. Plasma-enabled amorphous TiO_2 nanotubes as hydrophobic support for molecular sensing by SERS. *ACS Appl. Mater. Interfaces* **12**, 50721–50733 (2020).
225. Huh, Y. S., Chung, A. J. & Erickson, D. Surface enhanced Raman spectroscopy and its application to molecular and cellular analysis. *Microfluid. Nanofluid.* **6**, 285 (2009).
226. Hering, K. et al. SERS: A versatile tool in chemical and biochemical diagnostics. *Anal. Bioanal. Chem.* **390**, 113–124 (2008).
227. Yan, X., Li, P., Yang, L. & Liu, J. Time-dependent SERS spectra monitoring the dynamic adsorption behavior of bipyridine isomerides combined with bianalyte method. *Analyst* **141**, 5189–5194 (2016).
228. Garren, J. M., Abell, J. L. & Zhao, Y. Dynamic Rastering Surface-Enhanced Raman Scattering (SERS) Measurements on silver nanorod substrates. *Appl. Spectrosc.* **65**(7), 734–740 (2011).
229. Abell, J. L., Garren, J. M. & Zhao, Y. Dynamic rastering surface-enhanced Raman scattering (SERS) measurements on silver nanorod substrates. *Appl. Spectrosc.* **65**, 734–740 (2011).

230. Jen, Y.-J. et al. Tunable plasmonic resonances in TiN nanorod arrays. *Coatings* **9**, 863 (2019).
231. Goel, P., Kumar, S., Sarkar, J. & Singh, J. P. Mechanical strain induced tunable anisotropic wetting on buckled PDMS silver nanorods arrays. *ACS Appl. Mater. Interfaces* **7**, 8419–8426 (2015).
232. Namura, K., Kumar, S. & Suzuki, M. Application of self-assembled plasmonic structures for micro fluidic control and sensing. In: T. Tanaka & D. P. Tsai (eds.), *Plasmonics: Design, Materials, Fabrication, Characterization, and Applications XVIII* (vol. 11462), SPIE, p. 114621Q, 2020.
233. Rehman, I. U., Khan, R. S. & Rehman, S. Role of artificial intelligence and vibrational spectroscopy in cancer diagnostics. *Expert Rev. Mol. Diagn.* **20**, 749–755 (2020).
234. Shi, H. et al. Setting up a surface-enhanced Raman scattering database for artificial-intelligence-based label-free discrimination of tumor suppressor genes. *Anal. Chem.* **90**, 14216–14221 (2018).
235. Kumar, S. et al. Proof-of-concept: Smartphone- and cloud-based artificial intelligence quantitative analysis system (SCAISY) for SARS-CoV-2-specific IgG antibody lateral flow assays. *Biosensors (Basel)* **13**, 623 (2023).
236. Beeram, R., Banerjee, D., Narlagiri, L. M. & Soma, V. R. Machine learning for rapid quantification of trace analyte molecules using SERS and flexible plasmonic paper substrates. *Anal. Methods* **14**, 1788–1796 (2022).
237. Banaei, N. et al. Machine learning algorithms enhance the specificity of cancer biomarker detection using SERS-based immunoassays in microfluidic chips. *RSC Adv.* **9**, 1859–1868 (2019).
238. Schultz, Z. D. Rise of the machines: SERS instrumentation and machine learning enabling complex bioanalysis. *Spectroscopy* **35**, 9–10 (2020).
239. Beeram, R., Vepa, K. R. & Soma, V. R. Recent trends in SERS-based plasmonic sensors for disease diagnostics, biomolecules detection, and machine learning techniques. *Biosensors (Basel)* **13**, 328 (2023).
240. Barucci, A. et al. Label-free SERS detection of proteins based on machine learning classification of chemo-structural determinants. *Analyst* **146**, 674–682 (2021).

Chapter 2

Smart sensors

The enablers of Internet of Things

Arshdeep Singh

2.1 INTRODUCTION TO SMART SENSOR TECHNOLOGIES

The fundamental purpose of a sensor is to convert a physical characteristic, such as pressure, light, temperature, distance, and direction, into a suitable electrical signal that is proportional in nature. Historically, sensors were deemed sufficient for exhibiting attributes such as consistency, reliability, resolution, range, robustness, dependability, and awareness. However, in the contemporary era, sensors must be integrated with intelligent systems that require advanced features like remote working capability, low power consumption, and wireless connectivity stability (Kalsoom et al., 2020). The contemporary era has witnessed a growing demand for intelligent systems due to their capacity to gather, evaluate data, and make informed judgments. The adaptable nature of smart systems has led to their widespread adoption across several industries and personal products. One crucial attribute of intelligent systems is their capacity to make decisions based on data. In this regard, contemporary sensors assume a significant role by enabling efficient gathering and transmission of this data to the system (Dartmann et al., 2019). One intriguing growing field that utilizes intelligent sensors is the Internet of Things (IoT), which involves wireless networks and dispersed sensors to collect real-time data and provide desired outcomes through suitable processing (Maijer et al., 2014). Artificial intelligence (AI) plays a crucial role in enabling sensors to function as intelligent sensors, effectively utilizing them for various applications, for example, general environmental monitoring, factor-specific environmental monitoring, forecasting the weather, imagery from satellites, technologies based on remote sensing for risk event monitoring (e.g., landslides), health care, autonomous vehicles, and many more (Hamrita et al., 2005). In relation to the industry, there has been a significant surge in the use of intelligent devices within healthcare facilities and diagnostic centers. These gadgets are employed to assess and track diverse health issues of patients, both in a remote and physical manner. In contemporary scientific and research endeavors,

 DOI: 10.1201/9781003633884-2

the utilization of sophisticated sensors is important for achieving optimal performance (Schütze et al., 2018). The IoT facilitates the interconnection of both living and nonliving entities, resulting in transformative advancements. Various network media are utilized to establish connections between objects. The primary goal of the IoT is to enhance dynamism and convenience. The IoT domain has greatly expanded the quantity of intelligent objects and gadgets. It allows several objects (devices) to function as intelligent entities. The proliferation of IoT technologies has facilitated widespread connectivity and the provision of intelligent services (Ding et al., 2011). It is currently being extensively implemented in diverse intelligent applications and generates a broad spectrum of economic prospects. The IoT encompasses a range of components, such as cloud computing, mobile devices, virtualized environments, sensors, radio frequency identification (RFID), and AI. In the context of the IoT, gadgets have the capability to communicate information and offer a wide range of useful services. One of the many benefits that owners of Alexa devices and other IoT-enabled smart home appliances can enjoy is the ability to operate a wide variety of appliances, both inside and out, without the need to physically touch them (Iyengar et al., 2022). Devices such as Echo Spot and Alexa possess the capability to engage in audio-video communication, play music, stream videos, access news bulletins, display calendars, create to-do lists, manage traffic, access social network accounts such as Facebook images, monitor children remotely, control lighting and switches, among other functionalities. Amazon's latest "Tap to Alexa" functionality allows individuals with speech and hearing impairments to benefit. Users initiate typical "Alexa" commands by just tapping the screen without engaging in conversation. It is particularly beneficial for individuals with disabilities such as blindness and hearing loss (Vongsingthong and Smanchat, 2014). These gadgets are utilized in IoT-enabled environments for many purposes such as entertainment, music listening, exploring surrounding locations, setting timers, receiving news updates, weather updates, and many more. These gadgets are commercially accessible, both with and without screens. In the absence of a screen, devices solely offer auditory services, but devices equipped with a screen also give services accompanied by a visual display. The impact of the IoT smart environment on human existence is significant. Sensors play a crucial role in any intelligent application. The system identifies any alterations in physical or chemical properties, and subsequently utilizes the gathered data to automate the application or devices, thus enhancing their intelligence. The IoT encompasses a diverse range of sensors, devices, and nodes that possess the potential to establish communication (Meijer et al., 2014). Sensors in IoT applications facilitate the integration of the physical and digital realms through the utilization of fog computing.

2.2 OVERVIEW OF THE IoT

The IoT encompasses the complex interconnection of everyday objects or gadget. Sensors are devices that are integrated or affixed to objects, with the capability to perceive or interpret data, as well as retain the data for subsequent study. The utilization of various gadgets or objects, such as a basic smartphone that is portable, enables the streamlining, automation, monitoring, and analysis of our daily tasks in an effective manner. For a long time, the field of information technology has made significant use of the building blocks of the IoT, which include sensors, networked sensors, machine learning, and actual time positioning. To date, a universally accepted and formal definition of the notion of the IoT has not been established on a global scale (Vongsingthong and Smanchat, 2014). Nevertheless, other definitions exist that aid in comprehending the precise significance of IoT. The concept of the IoT emerged throughout the 1980s and gained global recognition in 1999. According to Vongsingthong and Smanchat (2014), the IoT is commonly regarded as the worldwide integration of physical things. These tangible items can be categorized into three types: activity-aware, policy-aware, and process-aware. Their awareness, interactivity, functionality, and representation in programming models determine their classification. There are three different ways to think about the IoT. Some of these viewpoints include those of sensing entities, the Internet as a unifying framework that links all things, and semantics, which concerns the communication protocols that regulate processing, from the point of view of the items or devices that make up the network (Ali et al., 2010). The IoT is widely recognized as a means of achieving universal computing across devices that possess distinct addressing schemes, enabling them to communicate and exchange data with one another (Gupta and Quamara, 2020). According to Tiwari and Singh (2016), the IoT has the potential to facilitate connectivity between items and individuals across many contexts, locations, and timeframes. This connectivity can be established through networks, pathways, services, or communication mechanisms (Patil et al., 2012). To comprehend the true objective of IoT and its impact on our lives in the coming years, let us examine a straightforward illustration of a smart home. Assuming an individual departs for work and inadvertently neglects to deactivate the lighting or cooling system within their residence, the utilization of smart home infrastructure enables them to remotely execute these actions from a distance. Likewise, upon his return from the workplace in the evening, upon entering his residence, the air conditioning system in the room autonomously adjusts ambient temperature based on the data obtained from the sensors positioned on his person. In addition, the refrigerator recommends food items such as an energy drink or fruits to restore energy levels, based on data collected from sensors placed on the individual's body. The music system adjusts the music based on the

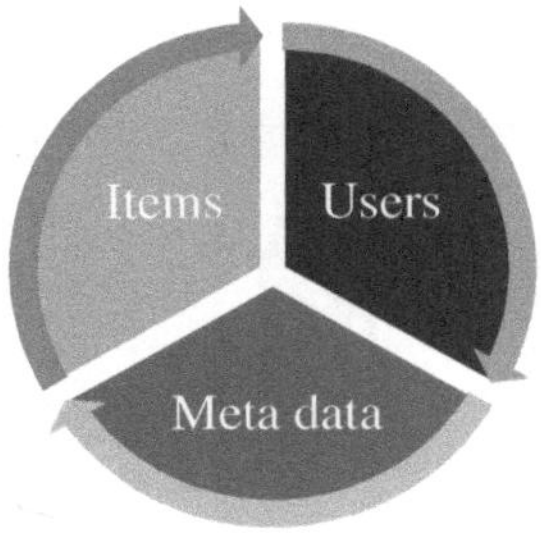

Figure 2.1 IoT system's components.

individual's mood, potentially playing rock or soft music. The lights automatically adjust to provide comfort or adjust to the individual's eyes, among other features. The potential outcomes of leveraging demonstrates that the IoT will undeniably enhance our lives to a significant degree. The IoT is a paradigm that encompasses various views (Malik and Om, 2018). The partnership known as CASAGRAS has extended the definition of this concept beyond the conventional focus on RFID. The primary objective of this consortium is to provide a worldwide perspective on delivering services that prioritize human needs. This is achieved by facilitating the automatic connection between things and computer systems, as well as their interconnections (Antora et al., 2023). IoT is prevalent in various industries and has the capability to provide a vast array of functionality. This will generate several job prospects and yield lucrative profits in terms of revenue. Therefore, it can be asserted that the IoT will experience significant expansion in the upcoming years (Figure 2.1).

2.2.1 IoT architectures

The comprehension of a system and its operational mechanisms necessitates the utilization of an architectural model (Figure 2.2). Numerous concepts for IoT architecture have been documented in previous scholarly works. An essential part of the IoT design is the hardware, which includes sensors and actuators; the middleware, which processes and transmits data; and the presentation, which is made to make it easy to use, understand, and carry around. Various components, including intelligent computer technologies, management services, authentication and authorization services, and others, are utilized by the application tier, which is the software interface layer that allows users to interact (Kumar and Patel, 2014). The network tier encompasses the communication protocols that facilitate networks, network infrastructure, and gateway operations for Internet connections

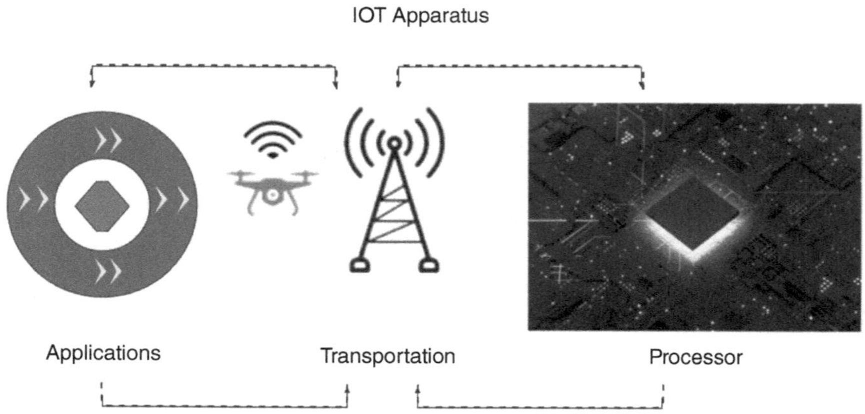

Figure 2.2 IoT applications/apparatus.

(Al-Qaseemi et al., 2016). The wireless sensor network (WSN) layer is an alternative term for this layer. This layer is the most important of the three since it serves as the architecture's primary functional unit, much like the brain or a computer's central processing unit (CPU). The physical tier encompasses several components such as data collectors or sensors, RFID technology, raw data, and real-time information. These components are effectively coordinated and collaborated to facilitate the transmission of data to the top processing layer (Jabraeil et al., 2020).

2.3 SMART SENSOR TECHNOLOGIES

2.3.1 Key Internet of Things technologies

RFID: The RFID system consists of radio frequency identifying tags and readers. Tags or labels are multiple in quantity, possess a specific location, and are affixed to items. A tag refers to a little microchip that incorporates an antenna. The utilization of an electromagnetic field facilitates the transmission or reception of data from an object via a tag. The electrical format of the tags stores the data, which can only be accessed by a reader when both tags fall within a predetermined range (Roberts, 2006). The data are transmitted by the reader through a signal, which is subsequently received and acknowledged by the antenna on the tag, which then proceeds to transmit the data. There can be one or more readers. The tags are available in three different configurations (Rajaraman, 2017). The passive reader, also known as the passive reader active tag, receives signals from tags that are powered by batteries. The transmission cover spans from 1 to 2,000 feet. The active reader passive tag is primarily utilized in many applications (Xiao et al., 2007).

WSN: It is a system comprising multiple nodes equipped with sensors and controllers. These nodes are responsible for sensing and monitoring data, as well as interacting with the environment. This phenomenon facilitates the establishment of connectedness among computing devices, persons, and their immediate environment. WSN plays a crucial role IoT. The central network tier of the IoT design relies on WSN for processing (Wang and Balasingham, 2010). Deploying sensors in a topology, detecting them, connecting them to the network, routing them, and transmitting information are crucial functions in a WSN (Nayyar and Singh, 2015).

Near-Field Communication (NFC) Technology: This is employed for the purpose of transmitting and relaying data in limited quantities between two devices while they are in proximity, towards the functionality of RFID. The wireless connection is established by the utilization of an integrated RFID reader in conjunction with a mobile phone. Low power is employed, and the ranges are limited. This form of radio communication operates by physical contact or by bringing devices into closer proximity to one another (Hamzah et al., 2019). An individual is required to be physically present at a retail establishment to make a payment. NFC serves as a fundamental technology for linking intelligent items in the IoT. Mobile NFC is an idea that has the capacity to effectively utilize basic technologies, like the evolution of mobile phones into payment gateways, particularly in the context of credit card transactions (Coskun et al., 2015).

Actuator: It is a specialized apparatus that is tasked with executing movements by utilizing a power source, such as hydraulic fluid or electric current. The device has the capability to generate a range of motions, including oscillatory, rotational, and basic linear motion. It has the capability to encompass a maximum distance of 30 feet (Hunter et al., 1991). The communication speed typically falls below 1 Mbps. Actuators find application in industries that involve the production or manipulation of mechanical components.

ZigBee: It is a versatile wireless networking technology created by the ZigBee Alliance in 2001 for short-range applications. ZigBee's main objective is to enhance the implementation of WSN. It has exceptional scalability and reliability, is cost-effective, and consumes minimal power. The device operates within a 100-m range and has a bandwidth of 250 kbps (Pan et al., 2007). Among the many fields that ZigBee may assist are smart energy, household automation, applications for manufacturing, and health care. The RF4CE specification, which is a reduced version of ZigBee, is exclusively employed for star topology. A remote health monitoring system can effectively utilize the ZigBee wireless network (Ramya et al., 2011).

Z-Wave: This is primarily designed for the purpose of facilitating applications for the automation of homes. Z-Wave was specifically engineered to prioritize a low data rate while maintaining high reliability, scalability, and low power consumption (Yassein et al., 2016). Z-Wave and ZigBee share some similarities, including their status as wireless protocols designed for low-power devices. A distinction can be observed between the two technologies. Z-Wave runs inside the frequency spectrum of 868 MHz, while ZigBee operates within the 2.4 GHz frequency band. Z-Wave has prioritized encryption in its software, while ZigBee employs the 128-bit AES encryption algorithm in its hardware.

Bluetooth LE: It, also known as Low Energy Bluetooth, is a wireless technology standard that eases the interchange of data between stationary and mobile devices within a limited distance. It is a new protocol designed specifically for applications requiring very low power consumption. The frequency range of Bluetooth is 2.4–2.485 GHz (Heydon and Hunn, 2012).

2.4 SENSOR APPLICATIONS FOR A SMARTER WORLD

Sensors are widely employed across various domains to establish an intelligent IoT ecosystem. Many scholars are currently engaged in the field of IoT sensors. This section provides an overview of the various applications of IoT sensors.

2.4.1 Smart cities

The mission of the smart city initiative is to enhance the existing urban infrastructure inside metropolitan areas. The incorporation of smart cities into urban development programs is a fundamental notion. Urban cities are equipped with a diverse array of technologies, including but not limited to the internet, wireless communication, infrared, Bluetooth, and Wi-Fi. These technologies exhibit a wide range of capabilities. The idea of a smart city is to maximize the utilization of public resources through the enhancement of service quality and cost reduction (Ramírez-Moreno et al., 2021). The primary objective of the IoT in urban areas is to facilitate convenient and distinctive access to public resources, thereby enhancing the efficiency and effectiveness of transportation monitoring, electricity distribution, and public area upkeep. The utilization of the smart city idea has the potential to enhance transparency and facilitate proactive measures by local authorities in addressing public demands. The sector of communication and digital technologies is seeing significant growth and advancement. The notion of a smart city, utilizing IoT equipment, is progressively advancing in intelligence (Hancke and Hancker Jr, 2013). The proliferation of digital devices, such as sensors, actuators, and smartphones, has led to substantial economic opportunities for the IoT. This is due to the ability of these devices to interconnect and communicate with one another via the Internet. The potential interest in smart cities is in the overviews of services that can be established in urban IoT. The objective is to improve the quality and range of services provided to the public, while also providing economic benefits to the city government by reducing operational expense (Antolín et al., 2017). The perception of a smart city encompasses several fundamental components, which are enumerated as follows:

Smart Parking: This aspect of a smart city deals with keeping an eye on parking cars sensibly by locating designated spots.

Construction Health: Through the monitoring of many factors such as material conditions and vibration levels of buildings, bridges, monuments, and roads, it is feasible to propose timely enhancements.

Smart Noise Management: Real-time sound monitoring can be implemented to proactively mitigate potential incidents in the bar area and densely populated areas.

Smart Traffic Management: The utilization of vehicle and congestion monitoring on various routes enables the provision of real-time recommendations and optimizations for driving and walking routes. Monitoring the levels of pedestrians and vehicles to improve the quality of walking and driving routes.

Enhanced Waste Management: The optimization of trash collection routes can be achieved through the detection of waste levels in various containers. Moreover, the integration of smart waste management with smart traffic management enables the provision of recommended routes for waste container vehicles to collect waste, based on the waste volume and traffic congestion.

Smart Lighting: Activating or deactivating streetlights based on weather conditions and specific needs (when a vehicle passes by) leads to significant energy conservation. The use of intelligent and weather-responsive lighting systems at elevated lighting locations along road edges.

Smart Roads: When we talk about "smart roads," we are referring to roadways that can monitor weather patterns and respond with alternate routes in the event of traffic jams, accidents, or other catastrophic events.

Internet of Energy (IoE): The IoE presents a novel framework for the distribution of power, storage of energy, monitoring of the grid, and communication. The IoE enables the transfer of energy units in a flexible manner, based on demand, timing, and location. Monitoring power usage will be conducted at several levels, ranging from individual devices at the local level to national and worldwide levels.

2.4.2 "Smart environment" or "smart earth"

The utilization of sensors to track environmental factors greatly contributes to national progress. This is due to the extremely complex deployment and maintenance requirements as well as the severe operational conditions (Ullo and Sinha, 2020). In the part that follows, we will discuss the use of sensors in making a smart environment:

1. **Detection of Forest Fires:** Monitoring of combustion gases and preventative fire conditions allows for the definition of alert zones and the timely delivery of warnings, for example, monitoring flue gases and fire prevention conditions to pinpoint danger zones.
2. **Pollution in the Air:** Industrial and vehicular sources of dangerous gases can have their emissions tracked and managed with the help of sensors. If addressed promptly, this has the potential to greatly reduce environmental impacts, for example, reducing emissions of carbon dioxide from power plants, cars, and other agricultural sources.
3. **Catastrophic Early Detection:** By keeping an eye on things like earth vibrations, soil density, soil wetness, and so on, potential landslides and earthquakes can be identified in plenty of time. Groundwater levels, Earth's density, and vibrations are being monitored for potentially hazardous trends in Earth's states.
4. **Protecting Wildlife:** Tracking collars locate and follow animals using GSM/GPS modules, then send their coordinates to a user's mobile phone via text message.
5. **Coast and Ocean Monitoring:** Utilizing a suite of sensors mounted on aircraft, ships, and satellites to keep the seas safe, track down missing fishing boats, identify potentially dangerous oil sources, and so on.
6. **Networks of Weather Stations:** Studying agricultural land's weather for the purpose of ice, drought, and air change prediction.

2.4.3 Intelligent water

Conserving water can be achieved by keeping tabs on the water's quality, its ground level, pollution level, and any container leaks. Water management is one area that could benefit from the usage of some of these smart sensors.

1. **Leaks in the Water System:** By monitoring the water pressure on exterior tanks and pipelines, leaks can be identified and addressed in a timely manner.
2. **Limiting the Amount of Pollution in the Water:** Sensors can be used to keep an eye on any hazardous or wasteful leaks or additions to the water.
3. A smart pool is one that uses sensors to track the pool's conditions and makes timely management decisions based on that data.
4. **River Flooding:** Keeping tabs on the water levels in rivers, dams, and reservoirs, as well as how they fluctuate over time, is a crucial responsibility.
5. **Water Quality:** An evaluation of whether or not water in naturally flowing rivers and oceans is suitable for human consumption and for all forms of life in a given area or period.
6. **Flooding:** Keeping an eye on how river, reservoir, and dam flow rates fluctuate.

7. **Supply Chain Control:** Keeping an eye on storage conditions while activities are in progress and using manufacture tracing for tracking purposes.
8. Water meter-to-IP network connection enables real-time data collection on water condition and use for water management purposes
9. Improving the quality of wine by keeping an eye on grape sugar levels and vine health through monitoring soil water retention and trunk width in vineyards.
10. **Golf Courses:** In dry regions, strategically placed sprinklers reduce vital green-water resources.
11. **Greenhouses:** Control the indoor temperature and humidity to boost fruit and vegetable production and brilliance.
12. **Water Quality Monitoring in the Field:** Reducing food spoilage by superior tracking and statistical management; managing crop fields through superior fertilization, irrigation, and electricity management; and continuously acquiring data.

2.4.4 Intelligent safety

Sensors can be placed in different places to make sure the environment is safe (Saifuzzaman et al., 2017). This section discusses a few examples of applications where sensors are crucial in ensuring security:

1. **Access Control with a Smart Perimeter:** By installing sensors and alarms around designated areas, unauthorized individuals can be promptly alerted when they attempt to enter restricted areas. With the ADXL 345 accelerometer, an intelligent intrusion system may be mounted to the door.
2. **Enhanced Explosive Detection:** Utilizing sensors, hazardous gas, radiation, and leak levels can be intelligently detected. In places where there is a risk of leaking, such as mines, factories, and other industrial settings, sensors can be fastened to containers or walls.

2.4.5 Smart transportation and mobility

Shipping Quality: Inspecting for vibrations, impacts, and openings of containers to ensure insurance coverage.

Pay via NFC: In order to pay for things like museum admission, gallery admission, or public transit, businesses can accept online payments using a link to a merchant bank or acquirer that is location- or time-dependent.

Object Location: Locating specific things in massive spaces like ports or repositories.

Keep an Eye on What's Going on with the Fleet's Cars and Assets: Controlling the route of valuables, pharmaceuticals, and dangerous goods.

Notice of containers emitting easily flammable compounds near others containing explosives in the event of a noncompatibility finding in storage.

Car Management: Companies that offer car-sharing services control the use of automobiles by connecting each one to the internet through a smartphone.

Automatic Vehicle Diagnostics: Controller area network (CAN) Bus data collection for real-time alert transmission of imminent dangers or advice to a driver.

2.4.6 Smart homes

By attaching sensors to various items around the house, we can create a smart environment. This allows us to do things like intelligently turn on and off appliances, detect intruders, and even measure the amount of ingredients used in cooking. We then receive alerts about these activities, which ultimately improves our lifestyle. Here are a few examples of smart home applications:

1. Intelligently turn appliances on and off at the touch of a button, making life easier and safer by reducing the likelihood of accidents and saving energy.
2. **Intrusion Detection Systems:** Sensors installed at access points can identify intruders and promptly notify the appropriate authorities. Since one can keep an eye on their house from any location, this greatly simplifies and secures daily life. This can be put into action when authentication is strictly enforced and only authorized individuals are permitted to enter (Belghith and Obaidat, 2016).
3. **Water and Energy Conservation:** With the use of sensors, it is possible to provide advice on how to manage the consumption of water and power.

2.4.7 Smart health

Human existence was profoundly impacted by the IoT. Patients can wear sensors at home, at work, or in hospitals as part of smart health or e-health. In the event that these sensors identify any abnormalities in a person's conditions, they will send out alerts to the appropriate parties. Patients and clinicians alike can benefit from smart bands worn in hospitals, which aid in patient treatment management. The employment of sensors allows for the monitoring of vital signs in the elderly, as well as the regulation of medical equipment, vaccinations, and organic materials (Formica and Schena, 2021). People might be warned of harmful habits, health issues, and even ultraviolet (UV) sun exposure. Medical alert systems assist people with disabilities or the elderly who live alone and are at risk of falling.

Smart tools also help in cold storage regulations for pharmaceuticals, vaccines, and organic components in pharmacy refrigerators. Hospitals and nursing homes use patient monitoring systems to keep tabs on residents' vitals. In the medical treatment for athletes, tracking vital signs at training facilities and competition camp is crucial. For this reason, there are fitness and health trackers available on the market that can determine your heart rate, weight, number of steps, and other vitals. Remote monitoring of patients with chronic illnesses is possible using patient surveillance tools that provide complete patient data. This aids in the management of these patients. A short hospital stay, cheap prices, and access to a smaller medical institution are all potential benefits. The patient's hand hygiene control system incorporates Bluetooth LE tags and RFID wristbands into a monitoring scheme that sends vibration alerts to warn the user when it is time to wash their hands. All of the data collected can be used to create statistics that can be used to monitor the health problems experienced by individual patients and medical professionals.

Determining when people should not be exposed to UV sunlight is possible by measuring it. Taking care of your teeth has never been easier than with a Bluetooth-connected dental brush and an app on your smartphone. While brushing your teeth, the brush can analyze your movements and send that data to your dentist or other healthcare provider, protecting your privacy.

2.5 INTERNET OF THINGS (IOT) CONNECTIVITY AND SMART SENSOR APPLICATION IN AGRICULTURE

Food production is fundamental to human subsistence. A number of steps have been implemented to boost the harvest yield. The agricultural loss is caused by pest infestations and hard environmental conditions. Increased agricultural output with less economic loss is possible with the use of cutting-edge technology such as improved sensors linked with the IoT. Research from the globe has proven that smart sensors connected to the internet can help keep an eye on soil moisture, temperature, humidity, and other growth-critical environmental variables (Rajak et al., 2023) and even measure greenhouse gases like carbon dioxide and methane. As an additional benefit of smart farming, farmers can now measure soil nitrogen concentrations, which helps them decide how much fertilizer to apply to their fields. For precise monitoring of insect attacks and related diseases in agricultural vegetation, some IoT-enabled equipment and unmanned aerial vehicles (UAVs) are helpful. High-resolution photographs are saved in databases on remote servers for later analysis in order to identify pest infestations in agricultural fields, even when we are in remote places. A smart greenhouse is a new kind of greenhouse that can grow plants with little to no human involvement. To improve crop productivity, it employs sensor-based continuous monitoring of environmental variables including light, humidity, temperature, and soil moisture (Sinha and Dhanalakshmi, 2022). Decisions can be made to conduct additional corrective actions to avert agricultural harm with the help of these automated sensors. A farm management system (FMS) integrates the IoT and AI into a unified plan. In order to improve the value and measure of agricultural produce, it provides a better tracking system for analyzing a wide range of physical, chemical, and biological parameters. With the use of real-time data processing facilities, it requires less labor. Thus, FMS can aid in the reduction of time and effort required to sustainably produce high-quality agricultural goods (Mat et al., 2018).

2.5.1 Remote sensing using smart sensors

A new breed of sensors known as "smart" sensors has emerged as state-of-the-art technology that can communicate wirelessly from afar, thanks to the explosion of innovation in computing, semiconductor manufacturing, and communication. They can communicate with one another and with the automated data processing network. Consequently, "smart" sensors excel at finding patterns in raw data, drawing conclusions about the interplay of different components, and establishing cause-and-effect relationships. In contrast to subsequent iterations of sensor networks, which were enhanced with intelligence to process calculations in addition to sensing input, earlier generations of sensor networks merely had sensing capabilities. The ability for sensors to communicate with each other is what gives them their "smart" characteristics; this allows for the creation of what is called a "smart sensor network" (Ullo and Sinha, 2021). The fundamental components of a smart sensor include a microcontroller, an Analog-to-digital converter (ADC), a communication link, memory, power, and one or more sensors. The three primary parts that comprise a smart sensor node are namely component parts, including a transducer, a memory core, CPU, and interface for the network. After picking up on the physical parameters, the physical transducer transforms them into an electrical signal. If the ADC is not present, the CPU will not be able to generate a digital value. The processor, often a microcontroller, processes the observed data by applying various signals and then transmits the results to the network. Traditional sensor networks consist of a single data processing node. Data are processed locally at the signal node and then communicated to the network as a result. Connected smart sensors with the ability to interpret and analyze data in real time can greatly enhance the efficiency of remote sensing. Agricultural remote sensing could make use of both active and passive sensors. Weather prediction, topography of landscapes, pest manifestation, soil quality, and monitoring of soil conditions are all areas where IoT-enabled smart sensors could be useful. In order to improve the accuracy of their operations, smart sensors rely on AI (Corsi, 2007). By providing more precise data in real time, light detection and ranging (LIDAR)-based sensor technology enables precision agriculture. Soil moisture, crop development, pest symptoms, plant growth status, and productivity may all be tracked in real time using this equipment. There is also broad data on drought and plant diseases provided. Data capture, processing, and analysis are just a few of the many areas that the IoT could affect. In addition, the IoT has been intensively studied for a number of environmental circumstances, including water contamination, pollution, soil quality evaluation, radiation status, and many more. Environmental research and agricultural applications can benefit from geospatial analysis that is tagged with environmental informatics and the IoT. "Smart environment" applications are increasingly relying on sensors that are linked to RFID devices for remote sensing capabilities (Rajaraman, 2017). An antenna, a tag, and a reader comprise an RFID system. With this system, the data stored in the tag can be retrieved with the help of radio transmission between the tag and the reader. The incorporation of sensing elements into existing RFID technology has great promise for the remote sensing of environmental stimuli like temperature and humidity, which play a vital role in agricultural techniques.

2.5.2 IoT and agriculture

Agricultural technological approaches enabled by the IoT also aid in evaluating soil erosion, crop quality, soil fertility, soil health, and the demand for fertilizer. Optical irrigation, monitoring crop growth at different stages, and seed quality are all areas it helps with. The processing of real-world data from remote sensing and the IoT can greatly benefit precision agriculture and forestry. Agronomic topological data can now be sensed using methods such as infrared thermography in conjunction with smart sensors. In agricultural fields, smart soil moisture sensors and the IoT are used to track the state of crops before and after harvesting (Farooq et al., 2019). Microbes have negative effects on people all around the world. Artificial pores based on microfluidics can detect the presence of microbes in agricultural foods. Lateral flow tests and assays are the mainstays of onsite pathogen detection. Interestingly, data collected by various sensors are interpreted using deep learning algorithms. Using IoT, ZigBee, and Arduino sensors, one may precisely measure the soil's condition, humidity, temperature, and crop types in a given region. Consequently, the IoT can evaluate data in order to increase agricultural and crop production in a planned manner. The IoT and UAVs are the backbone of smart farming, allowing for more efficient management of resources with less human intervention and higher yields (Farooq and Akram, 2021). Smart farming benefits even more from the more precise real-time data provided by flying IoT. There has been a worldwide focus on microprocessor development and deployment due to the potential benefits of automated intelligent regulation of IoT services. Data retrieved from satellites and GPS systems allows many industrialized nations to conduct precise, real-time monitoring of field cultivation. Farmers can improve their planting and crop management tactics with the use of agricultural IoTs coupled with expert systems. According to Muangprathub et al. (2019), there are significant applications for electromechanical sensors, biosensors, and physical property sensors in the agricultural sector. To demonstrate smart farming, a number of organizations have set up innovative IoT systems that use cloud computing and Li-Fi. The term "Li-Fi" describes a type of wireless data exposure that is very concentrated in a small area. In terms of throughput, efficiency, accessibility, and security, Li-Fi outshines Wi-Fi. Activities such as spraying, weeding, and moisture detection are the primary goals of this intelligent agricultural operation. Additionally, it incorporates smart warehouse monitoring by guaranteeing the maintenance of temperature and humidity in the stockroom. Last but not least, a dependable and transparent application of agricultural operations requires perfect irrigation with exquisite management. To achieve all of these goals, smart modules such as Li-Fi, ZigBee, or edging sensors take action. Prudent water use is of paramount importance in light of the fact that water is becoming scarcer by the day. The smart irrigation decision support system (SIDSS) is an innovative tool with promising future use in agricultural water management. Two machine learning methods, namely ANFIS and partial least squares regression, have been suggested for the purpose of SIDSS.

2.5.3 IoT-based crop monitoring

Technological advancements made possible by the IoT allow farmers to keep a close eye on the development and condition of their crops. Furthermore, it aids farmers in real-time evaluation of pest attacks and plant illnesses. Owing to the IoT tagged sensors, researchers and farmers may intelligently control agricultural cultivation, fertilizer application, irrigation, and plant surroundings in real time. IoT smart crop monitoring systems rely on field-placed sensors to collect data on a variety of environmental factors, such as soil moisture, temperature, humidity, and nutrient levels (Kim et al., 2020). Connectivity between these sensors allows for continuous data transfer to a central server in the cloud. The data collected from these sensors are subsequently analyzed using machine learning and other data analytics techniques to learn about the crop's vitality, pace of growth, and prospective yield. This data is useful for farmers because it allows them to make informed decisions about irrigation, pest control, and harvesting. Using wireless sensor technology, administrators and farmers can be notified when equipment fails and begin debugging immediately. Reduced energy consumption, increased data processing, and improved actuation are further potential benefits of an automated repair tool. There has been a recent uptick in the use of Unmanned Aerial Systems (UAS) that are IoT-enabled in agricultural workflows for processing and data storage

(Farooq et al., 2020). The purpose of the CC2430 study was to assess several agricultural parameters through the use of an intelligent irrigation system. The system's utility in reducing water usage by crops was 22.6% higher than that of conventional irrigation techniques, and its installation cost was 44.8% lower than that of comparable foreign products. To make visual and environmental data collection even more convenient, UAS can be connected to smartphones, smartwatches, and laptops. Using devices based on the IoT, we are able to accurately conduct multipoint assessment systems that include field surveillance, data collecting, and monitoring. The IoT has expanded the digital system's usefulness in pest and disease management in greenhouses by allowing for computerized data processing and controlled oversight. IoT devices allow farmers to quickly and accurately track pests, allowing them to take the necessary preventative measures (Mohanraj et al., 2016; Shenoy and Pingle, 2016).

2.5.4 Application of IoT on weather sensing

The IoT has recently gained popularity as a means to detect a variety of meteorological characteristics, including soil moisture, temperature, and humidity. IoT in conjunction with wireless networks and smart sensors allows for more precise farming in vineyards and wineries by providing real-time data on environmental parameters. IoT in conjunction with sensors will immediately notify the administrator to take the required measures in the event that critical environmental elements are altered beyond a predetermined threshold. For real-time crop growth monitoring, the smart system additionally includes CO_2 concentration, light, humidity, and temperature (Tiwari et al., 2020). Interestingly, smart sensors connected to the internet are becoming more common in cucumber farming as a means to detect cryogenic hazards using environmental indicators in real time. This system works by using IoT sensors to monitor temperature; when the temperature drops below a certain point, an alert message is sent to the administrator (Sharma and Prakash, 2021). In this way, the administrator can prevent cold shock in the plants by turning on heating equipment. By integrating IoT with third-party weather forecasting systems, we can stay informed about impending weather changes and prepare for potentially harmful climate change (Pauzi and Hasan, 2020).

2.5.5 IoT-based soil property monitoring

Using smart sensors and IoT-coupled technologies, farmers may keep tabs on essential soil parameters for plant development and growth. Soil moisture, temperature, and nutrient content may all be detected by the IoT, which can then be relayed to farmers. To ensure that farmers take adequate precautions against crop diseases and pest attacks, they can calibrate and remotely access this data. Consequently, smart farming could be useful for farmers.

Additionally, farmers can use IoT to remotely monitor soil pH and rhizosphere zone multiparameter, allowing them to take preventative actions. It is also possible to track how plants develop in different kinds of soil. As a result, smart IoT systems give farmers real-time data on soil moisture, temperature, and condition, which boosts yield and quality (GASANA, 2022).

2.5.6 Sensors for smart farming

Multiple sensors are starting to make a splash in the smart farming space. These sensors have the potential to automate a number of agricultural processes, including harvesting, environmental monitoring, and yield measurement. "Smart sensors" are sensors that have integrated circuits. Subashini et al. (2018) found that smart sensors can more accurately capture and store a variety of environmental data and related information in agricultural setups on Single shot detector (SSDs). Microprocessors are used to interpret and analyze these data sets. IoT components include smart sensors that transfer data via the internet. From a few hundred to thousands of nodes linked to central sensor hubs make up this wireless actuator network system (Terence and Purushothaman, 2020). So, to enable remote monitoring of many agricultural parameters, a "smart sensor" combines a sensor device with microprocessors and wireless communication technologies. To get the most out of smart sensors, you can pair them with additional components like transducers, amplifiers, digital-to-analog converters, and analog filters. Below, we have covered a few sensors that can be quite helpful (Figure 2.3).

2.5.6.1 Acoustic-based sensors

When an object is in motion, these sensors can pick up on the vibrations the object returns. When moving about, foraging, or mating, many different kinds of insects make noises. Portable accelerometer and microphone systems are just two examples of the acoustic instruments that can pick up on these kinds of noises. Researchers from all over the world can use the data collected by the acoustic devices to improve automated acoustic sensors for species-level pest detection (Bai et al., 2020). When used in agricultural practices, acoustic sensors can get covered in dirt and dust, but that does not stop them from working efficiently and giving more precise information about the objects they are detecting. The increased sensitivity, reduced maintenance costs, and improved accuracy of acoustic sensors are their main selling points. Agricultural fields often employ acoustic-based sensors for the detection and monitoring of insect infestations. Furthermore, when planning irrigation schedules for automated irrigation systems, it is crucial to estimate the volume of water in water sources. According to Hoummady et al. (1997), mechanical flow meters are being progressively replaced with ultrasound flow meters that use ultrasound distance sensors that are

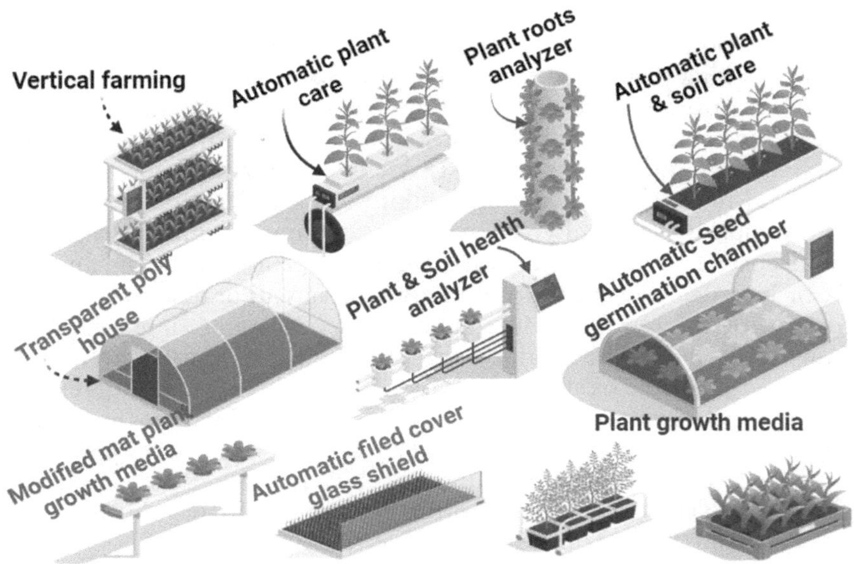

Figure 2.3 Sensors for smart farming.

enabled by the IoT. With the advent of underwater ultrasonic scanning, it is now possible to track the development of aquatic plants, which holds great promise for pinpointing when to harvest and how quickly they are growing. In addition, it is essential for fruit harvesting. Estimating when fruit is ripe is an interesting use of acoustics in farming.

2.5.6.2 *Electromagnetic sensors*

Agricultural soil and contaminants can be sensed by electromagnetic sensors. In addition, it is useful for mapping the topological characteristics of farmland. Various objects emit electromagnetic waves, and electromagnetic sensors can detect this spectrum. This method makes use of electrical circuits that record the electric impulses that accumulate or conduct through soil (Baum, 1986). Soil characteristics, conditions, and processes, as well as their spatiotemporal variability, can be studied using electromagnetic induction (EMI) (Passarao et al., 2006). This was made possible by the high-resolution, nondestructive repeated measurements, potential reduction in manpower requirements, and extensive spatial coverage offered by Ground Penetrating Radar (GPR) and EMI techniques. Soil electrical conductivity measurements obtained by sensor and laboratory methods were found to be very similar. In agricultural settings, the results suggest that electromagnetic sensors might be useful for precisely detecting a number of physicochemical soil properties

2.5.6.3 Electrochemical sensors

These low-power, environmentally friendly sensors are small and lightweight, making them ideal for use in agricultural settings. The efficiency of these sensors allows for better real-time monitoring of plant growth, illnesses, and environmental pollutants. In addition, electrochemical sensors have relatively little effects on the environment. For biochemical parameters vital to agricultural output, multiple sensors have been evaluated. One example is the use of graphene oxide-containing humidity sensors for the purpose of monitoring the water content of plant leaves (Kim and Lee, 2022). In order to detect harmful gasses in agricultural settings, a NO_2 sensor made of silver and reduced graphene oxide can be employed. As an example, a completely bendable device was built with sensing layers made of reduced graphene oxide coated with Silver nano particles (AgNPs) and conducting electrodes made of metallic single-walled carbon nanotubes. Even at ambient temperature, the sensor picked up levels of 0.2 ppm NO_2. In comparison to the sample without AgNP decoration and the sample based on metal electrodes, the sensitivity was approximately thirteen times higher and thirty-three times higher, respectively. Soil volatile organic molecule detection is aided by specific sensors (Chugh et al., 2022). The usual components of these types of sensors include gold nanoparticles, reduced graphene oxide, graphite, and carbon nanotubes (Li et al., 2010). It is worth noting that thirteen distinct plant volatiles were detected with a classification accuracy of over 97% using a multiplexed sensor array. Soil nitrogen and plant sap can be detected by some of the electrochemical sensors. Hence, electrochemical sensors have great promise for sensing a variety of factors in agricultural fields that could contribute to improved agricultural output (Ali et al., 2020).

2.5.6.4 Light detection and ranging (LIDAR)

LIDAR technologies use light to determine how far away an object is from a sensor. In LIDAR, the distance between the light source and the target can be precisely estimated since light always travels at the same speed. Using a sequence of collisions detected by light pulses emitted at regular intervals, LIDAR is able to construct a map of its immediate surroundings. A more modern LIDAR system will have fewer pulses that boost its performance. According to Long and McCallum (2013), other performance features encompass the operational range, predicted error, and scanning frequency. A file called "cloud point" is used to hold the coordinates that are obtained from light collision. This file aids in the construction of 3D space. A wide range of agricultural uses have made use of LIDAR technology. For instance, it tracks the structure of vegetation and the ripening of fruits. The topography and agricultural landscaping are also measured using LIDAR sensors. Using LIDAR technology, tree structural traits including canopy volume and leaf area index may be sensed (Rivera et al., 2023). This new technique is also used to estimate crop biomass, characterize phenotypes, and track development. It is also possible to locate complexes of agricultural terraces by incorporating slope contrast mapping. The use of FX6 LIDAR to detect Nippon Signal has the potential to revolutionize plant and ground segmentation (Debnath et al., 2023). In order for autonomous agricultural robots to carry out tasks like mapping, navigation, and localization, it is an essential prerequisite. Consequently, LIDAR may prove to be quite useful for mapping agricultural fields and launching automated farming techniques.

2.5.6.5 *Optical sensors*

A variety of optical sensors may find use in farming. Their capacity to pick up light of different wavelengths is the foundation of these sensors. A light source emits light with a certain wavelength in order to collide with the item of interest. The optical sensor records reflectant data, which is a text file of information about the reflected light. In order to analyze nutrient uptake, Freidenreich et al. (2019) used SPAD and GreenSeekerTM, two nondestructive optical sensors. We looked at things like total leaf carbon:nitrogen levels, soil leachate, and the normalized difference vegetation index. The authors believe that optical sensors could be a useful tool for estimating how much fertilizer plants need. For accurate harvest biomass forecast and canopy expansion/senescence monitoring, proximal optical sensors like SPAD-502, Green Seeker, and Canopeo App were employed. Here, SPAD-502 and Green Seeker optical sensors were able to successfully forecast crop biomass at harvest with coefficients of determination of 0.68 and 0.82, respectively, over the growing season. It has been reported that optical coherence tomography could be useful for monitoring seed germination. While planting wheat, the Green SeekerTM optical sensor can help with nitrogen fertilizer management for increased yield (Costa et al., 2023). The relative error percentage between observed and predicted values for Green Seeker was 5.80%, while for SPAD-502 it was 4.12%. Optical sensors combined with spectroscopy were shown to be very effective in identifying green weed-infected cells in a study that used a tolerance threshold. The value was similar to 90.5% and 91.2% of the two reference methods utilized, and the total agreement was 90.9%.

2.5.6.6 *Eddy covariance (EC) based sensors*

Sensors based on EC are used to quantify evapotranspiration and water use by crops. These sensors record sensible heat fluxes and latent heat fluxes. Measuring greenhouse gasses is another area where EC shines. By combining EC with remote sensing models, we can get more accurate estimates of crop gross and net primary productivity and greenhouse gas balance.

Ammonia surface-atmosphere exchange is measured using EC based on quantum cascade lasers. Researchers in the field used EC to monitor soil heat flux and evapotranspiration while watering with sprinklers (Kang and Cho, 2021).

2.5.6.7 Mechanical and mass flow sensors

Soil physical and chemical variables cause agricultural land crop production fluctuation. The compaction of soil particles in a certain location is reflected in soil strength. Reduced agricultural yields are associated with soils that are more compact. Soil compaction and mechanical resistance can be measured using mechanical sensors. Depending on the mechanical characteristics of the ground, this data might be useful for tailoring the irrigation method. One type of mechanical sensor is the Honeywell FSG15N1A (Jadhav et al., 2014). Honeywell FSG15N1A detects the pulling power of plant roots as they take in water. Researchers measured the density of soil at a specific location using both horizontal and vertical sensors. The depth to which the coulter penetrated the topsoil layer was measured using an ultrasonic distance sensor in this integrated approach. In order to detect changes in topsoil strength with respect to depth, the results showed that the integrated method was helpful. Soil mechanical resistance mapping in agricultural areas could benefit from this. The use of mass flow sensors opens the door to the possibility of automated crop yield monitoring. One study used a portable grain mass flow test setup to check how well mass flow sensors worked. The results of the analysis showed that the grain elevator's slope can influence the sensors' calibration, which in turn can change the system's accuracy (Loghavi, 2013). A ten-pitch adjustment to the calibration greatly changed the mass flow sensor's accuracy. Notably, when operated with a calibrated flow rate, the mass flow sensor achieved an impressively high coefficient of determination of 0.99, demonstrating its impressive precision. When comparing agricultural yields, impact-based grain flow sensors are commonly utilized. Even yet, it is possible for the overwhelming natural vibrations to cause erroneous indications. A dynamic compensation technique could be useful for improving the sensors' accuracy by reducing the overshoot and natural vibration signals.

2.5.6.8 Versatile and embedded sensors

Wearable and flexible sensors: More and more people are turning to these types of sensors because of the real-time physiological data they can supply about plants. For improved signal detection, these sensors can be applied directly to plant surfaces. The intricate surfaces of plant parts are taken into account while designing flexible sensors. Porous materials used to make leaf sensors do not impede plants' ability to breathe through their leaves (Yin et al., 2021). Another type of sensor is the flexible fruit sensor, which can quickly transform a flat surface into a three-dimensional structure and, as a result, operate efficiently on a variety of surfaces. Specifically, these sensors are engineered to document variations in microenvironmental temperature. A fruit-growing sensor that is both stretchy and flexible, is created by combining graphite powder with a chitosan solution in specific ratios. These sensors are rubber-coated to keep them dry. Cucumber fruits had these strain sensors—which can detect strains of up to 60%—written directly onto them so that they could quantify elongation. The primary goal of developing wearable sensors is to track people's vital signs. Nevertheless, these sensors have only just begun to be used in agricultural applications (Qu et al., 2021). To examine changes in microenvironmental parameters like temperature, wearable sensors can be directly placed on different sections of plants (stems, leaves, etc.). This allows for the detection of sick states and other forms of biotic and abiotic stress. In order to measure stem elongation, some stretchable sensors have been used. These sensors consist of polydimethylsiloxane stretching material that is coated with Ti/Au metal film. Patil et al. (2015) used stretchy latex sensors covered in graphite and carbon nanotube ink to track the circumferential elongation of *Cucurbita pepo* fruits. The elongation may be accurately measured at the micrometer scale using this sensor. This method has the potential to greatly improve the accuracy of plant growth measurements due to its streamlined process, reduced implementation cost, increased speed, and nanoscale sensitivity. Stretchability of wearable sensors composed of gallium-based liquid alloy can reach 200%, thanks to its high fluidity and electrical conductivity. Roses and bean sprouts are among the plants that have been evaluated for their ability to adopt unusual shapes using these sensors.

2.5.7 Pest monitoring in smart farming

Several agricultural pests can be repelled by devices that emit ultrasonic sound waves. For example, Dutta et al. (2020) devised a device that emits powerful ultrasonic sound waves to ward off pests. Insects, rodents, and other pests are scared away by these waves. Consequently, these tools are a great substitute for conventional pesticides, which have many negative effects on helpful creatures. Another factor that can lead to increased use of ultrasonic equipment is accidental contact to chemical insect repellents. There is no harm to humans or the environment caused by electronic insect repellants, and they are inexpensive. These devices primarily work by interfering with the pests' auditory systems. Pest detection is much enhanced by specific image processing methods (Koshy et al., 2018). Utilizing UAV technology, spectral cameras may take macro and micro photographs of the farmland with remarkable detail. To manage insect populations, you can set up automated traps that use cameras. In an effort to manage the codling moth, *Cydia pomonella* L., in apple orchards, Guarnieri et al. (2011) converted a commercially available trap into an automated one that could collect and interpret data. A delta-shaped trap outfitted with a GPRS, solar panel, battery, charging device, and high-resolution camera was devised by Preti et al. (2021). The *Lobesia botrana* Denis, a grapevine moth native to Europe, was captured in this automated trap. In order to keep an eye on the *Ceratitis capitata* fruit fly, a Mediterranean species, Shaked et al. (2018) developed sticky delta traps. Fruit flies, scientifically known as *Bactrocera oleae*, were captured using electronic McPhail traps that included buckets and cameras. The fruit flies that were caught were photographed automatically and saved on a server. Entomologists achieved an accuracy rate of over 88% when they used the photographed fruit flies to confirm their identification. A system to monitor the agricultural environment can also be set up to employ Field Programmable Gate Array (FPGA). The many sensors, screens, wireless protocols, microcontroller units, field-based field-programmable gate arrays, and serial protocols that make up such a monitoring system are all standard. Sensors measure important environmental variables like temperature, humidity, and soil moisture, and the FPGA displays this data on the screen. A study used a FPGA-based image processing (FIP) device, and it successfully solved the technical problems with traditional imaging systems' processing speed and image resolution (Preti et al., 2021). With a Lin's concordance correlation coefficient of 0.99 for the DLSR and 0.91 for the web camera reference system, the FIP system was found to provide great accuracy in this case. The superior precision of the FIP system compared to the web camera reference method is, thus, self-evident.

Challenges and Potential Prospects: Worldwide investigation on the IoT and smart farming using smart sensors has yielded optimistic outcomes. Many limited agricultural farms have begun to utilize IoT technologies. That being said, widespread use of this technology has not yet occurred. The financial consequences of implementing and integrating IoT-enabled sensors and accessories across a large area of agricultural land are one of the main barriers. Furthermore, there exists ambiguity regarding the expense of execution and the level of financial advantage achieved. The primary expenses associated with the application of IoT-enabled technology encompass the procurement of hardware, the deployment of software, and the ongoing operation of the system. Additional expenses may be incurred for energy consumption, system maintenance, service registration, and labor costs related with operating the integrated hardware devices and their corresponding software. There is a pressing need for a comprehensive global initiative to enhance the technological literacy of farmers, in order to facilitate the widespread implementation of IoT technologies. Economic policies that help farmers make better use of the IoT in agricultural settings are an absolute need from policymakers in the government.

It is imperative for government policymakers to provide economic policies that facilitate the effective and enhanced use of the IoT by farmers in agricultural areas. The issue of data privacy and security poses a significant challenge that can have adverse effects on the widespread adoption of IoT and smart devices. Attackers can manipulate data stored on cloud servers to disrupt automated farming procedures in agricultural fields. Hence, the challenges pertaining to data security in the IoT constitute significant factors contributing to the sluggish uptake of smart farming technology. The implementation of a robust encryption system is necessary in the realm of smart farming to safeguard critical data and digital systems from potential cyber threats on a global scale. The integration of cryptography with robust keys has the potential to mitigate cyber-attacks on cloud servers. Another approach that can yield a dependable result involves implementing a multiparty computation system that is connected with homomorphic encryption or paired with the blockchain technology. A significant issue with IoT devices is their outdoor placement, which exposes them to adverse climatic conditions such as heavy rainfall, dust, wind, temperature, and so on. The presence of unfavorable climatic conditions could lead to unforeseen mechanical malfunction of the advanced equipment

Summary and Future Prospects: The future of IoT-enabled smart farming holds promising prospects. IoT-enabled devices have the potential to facilitate the measurement of water and nitrogen levels in soil. Furthermore, the monitoring of evapotranspiration rates can be enhanced through the assessment of CO_2 levels in agricultural lands, hence facilitating improved surveillance of crop health. In addition, the mitigation of pest infestations can be achieved by implementing IoT-enabled traps that are equipped with high-resolution cameras and various supplementary devices. Although IoT-based smart sensors offer numerous benefits, its widespread implementation in agriculture faces significant obstacles such as the high costs associated with acquiring and maintaining advanced hardware and software. Furthermore, there is a dearth of awareness among farmers dwelling in rural areas pertaining to the utilization of IoT devices. Ultimately, cyber assailants have the potential to disrupt automated smart agricultural systems by causing harm to the cloud servers that house crucial data.

REFERENCES

Al-Qaseemi, S. A., Almulhim, H. A., Almulhim, M. F., & Chaudhry, S. R. (2016, December). IoT architecture challenges and issues: lack of standardization. In: *2016 Future Technologies Conference (FTC)* (pp. 731–738). IEEE, San Francisco, CA.

Ali, M. A., Dong, L., Dhau, J., Khosla, A., & Kaushik, A. (2020). Perspective—electrochemical sensors for soil quality assessment. *Journal of the Electrochemical Society*, *167*(3), 037550.

Ali, Z. H., Ali, H. A., & Badawy, M. M. (2010). IoT: definitions, challenges and recent research directions. *International Journal of Computer Applications*, *128*(1), 37–47.

Antolín, D., Medrano, N., Calvo, B., & Pérez, F. (2017). A wearable wireless sensor network for indoor smart environment monitoring in safety applications. *Sensors*, *17*(2), 365.

Antora, S. S., Chang, Y. K., Nguyen-Quang, T., & Heung, B. (2023). Development and assessment of a field-programmable gate array (FPGA)-based image processing (FIP) system for agricultural field monitoring applications. *AgriEngineering*, *5*(2), 886–904.

Bai, Y., Lu, L., Cheng, J., Liu, J., Chen, Y., & Yu, J. (2020). Acoustic-based sensing and applications: a survey. *Computer Networks*, *181*, 107447.

Baum, C. E. (1986). Electromagnetic sensors and measurement techniques. In: J. E. Thompson & L. H. Lussen (eds.), *Fast Electrical and Optical Measurements: Volume I—Current and Voltage Measurements/Volume II—Optical Measurements* (pp. 73–144).NATO AI Series.

Belghith, A., & Obaidat, M. S. (2016). Wireless sensor networks applications to smart homes and cities. In: M. S. Obaidat & P. Nicopolitidis (eds.), *Smart Cities and Homes* (pp. 17–40). Morgan Kaufmann, Burlington, MA.

Chugh, B., Thakur, S., Singh, A. K., Joany, R. M., Rajendran, S., & Nguyen, T. A. (2022). Electrochemical sensors for agricultural application. In: A. Denizli, T. A. Nguyen, G. Yasin, S. Rajendran, & A. K. Nadda (eds.), *Nanosensors for Smart Agriculture* (pp. 147–164). Elsevier, Amsterdam.

Corsi, C. (2007). Smart sensors. *Infrared Physics & Technology*, *49*(3), 192–197.

Coskun, V., Ozdenizci, B., & Ok, K. (2015). The survey on near field communication. *Sensors*, *15*(6), 13348–13405.

Costa, B. N. S., Tucker, D. A., & Khoddamzadeh, A. A. (2023). Precision horticulture: application of optical sensor technology for nitrogen monitoring status in cocoplum, a native landscaping plant. *Plants*, *12*(4), 760.

Dartmann, G., Song, H., & Schmeink, A. (Eds.). (2019). *Big Data Analytics for Cyber-Physical Systems: Machine Learning for the Internet of Things*. Elsevier.

Debnath, S., Paul, M., & Debnath, T. (2023). Applications of LiDAR in agriculture and future research directions. *Journal of Imaging*, *9*(3), 57.

Ding, D., Cooper, R. A., Pasquina, P. F., & Fici-Pasquina, L. (2011). Sensor technology for smart homes. *Maturitas*, *69*(2), 131–136.

Dutta, J., Dutta, J., & Gogoi, S. (2020). Smart farming: an opportunity for efficient monitoring and detection of pests and diseases. *Journal of Entomology and Zoology Studies*, *8*, 2352–2359.

Farooq, M. S., & Akram, S. (2021). IoT in agriculture: challenges and opportunities. *Journal of Agricultural Research, 59*(1), 03681157.

Farooq, M. S., Riaz, S., Abid, A., Abid, K., & Naeem, M. A. (2019). A survey on the role of IoT in agriculture for the implementation of smart farming. *IEEE Access*, *7*, 156237–156271.

Farooq, M. S., Riaz, S., Abid, A., Umer, T., & Zikria, Y. B. (2020). Role of IoT technology in agriculture: a systematic literature review. *Electronics*, *9*(2), 319.

Formica, D., & Schena, E. (2021). Smart sensors for healthcare and medical applications. *Sensors*, *21*(2), 543.

Freidenreich, A., Barraza, G., Jayachandran, K., & Khoddamzadeh, A. A. (2019). Precision agriculture application for sustainable nitrogen management of Justicia brandegeana using optical sensor technology. *Agriculture*, *9*(5), 98.

Gasana, J. M. (2022). *IoT based monitoring system of soil properties* (Doctoral dissertation, College of Science and Technology).

Guarnieri, A., Maini, S., Molari, G., & Rondelli, V. (2011). Automatic trap for moth detection in integrated pest management. *Bulletin of Insectology*, *64*(2), 247–251.

Gupta, B. B., & Quamara, M. (2020). An overview of IoT: architectural aspects, challenges, and protocols. *Concurrency and Computation: Practice and Experience*, *32*(21), e4946.

Hamrita, T. K., Kaluskar, N. P., & Wolfe, K. L. (2005, October). Advances in smart sensor technology. In: *Fourtieth IAS Annual Meeting: Conference Record of the 2005 Industry Applications Conference, 2005.* (Vol. 3, pp. 2059–2062). IEEE.

Hamzah, M. L., Desnelita, Y., Purwati, A. A., Rusilawati, E., Kasman, R., & Rizal, F. (2019). A review of near field communication technology in several areas. *Revista Espacios*, *40*(32).

Hancke, G. P., & Hancke Jr, G. P. (2013). The role of advanced sensing in smart cities. *Sensors*, *13*(1), 393–425.

Heydon, R., & Hunn, N. (2012). Bluetooth low energy. *CSR Presentation, Bluetooth SIG* https://www.bluetooth.org/DocMan/handlers/DownloadDoc.ashx.

Hoummady, M., Campitelli, A., & Wlodarski, W. (1997). Acoustic wave sensors: design, sensing mechanisms and applications. *Smart Materials and Structures*, *6*(6), 647.

Hunter, I. W., Hollerbach, J. M., & Ballantyne, J. (1991). A comparative analysis of actuator technologies for robotics. *Robotics Review*, 2, 299–342.

Iyengar, K. P., Kariya, A. D., Botchu, R., Jain, V. K., & Vaishya, R. (2022). Significant capabilities of SMART sensor technology and their applications for Industry 4.0 in trauma and orthopaedics. *Sensors International*, *3*, 100163.

Jabraeil Jamali, M. A., Bahrami, B., Heidari, A., Allahverdizadeh, P., Norouzi, F., Jabraeil Jamali, M. A., ... & Norouzi, F. (2020). IoT architecture. In: *Towards the Internet of Things: Architectures, Security, and Applications* (pp. 9–31). Springer, EAI.

Jadhav, U., Khot, L. R., Ehsani, R., Jagdale, V., & Schueller, J. K. (2014). Volumetric mass flow sensor for citrus mechanical harvesting machines. *Computers and Electronics in Agriculture*, *101*, 93–101.

Kalsoom, T., Ramzan, N., Ahmed, S., & Ur-Rehman, M. (2020). Advances in sensor technologies in the era of smart factory and industry 4.0. *Sensors*, *20*(23), 6783.

Kang, M., & Cho, S. (2021). Progress in water and energy flux studies in Asia: a review focused on eddy covariance measurements. *Journal of Agricultural Meteorology*, *77*(1), 2–23.

Kim, M. Y., & Lee, K. H. (2022). Electrochemical sensors for sustainable precision agriculture: a review. *Frontiers in Chemistry*, *10*, 848320.

Kim, W. S., Lee, W. S., & Kim, Y. J. (2020). A review of the applications of the IoT for agricultural automation. *Journal of Biosystems Engineering*, *45*, 385–400.

Koshy, S. S., Sunnam, V. S., Rajgarhia, P., Chinnusamy, K., Ravulapalli, D. P., & Chunduri, S. (2018). Application of the IoT for smart farming: a case study on groundnut and castor pest and disease forewarning. *CSI Transactions on ICT*, *6*, 311–318.

Kumar, J. S., & Patel, D. R. (2014). A survey on internet of things: security and privacy issues. *International Journal of Computer Applications*, *90*(11), 20–27.

Li, S., Simonian, A., & Chin, B. A. (2010). Sensors for agriculture and the food industry. *The Electrochemical Society Interface*, *19*(4), 41.

Loghavi, S. N. M. (2013). Design, development and evaluation of a mass flow sensor for grain combine harvesters. *Journal of Agricultural Machinery*, *3*(2), 10.

Long, D. S., & McCallum, J. D. (2013). Mapping straw yield using on-combine light detection and ranging (lidar). *International Journal of Remote Sensing*, *34*(17), 6121–6134.

Malik, A., & Om, H. (2018). Cloud computing and internet of things integration: architecture, applications, issues, and challenges. In: W. Rivera (ed.), *Sustainable Cloud and Energy Services: Principles and Practice* (pp. 1–24). Springer.

Mat, I., Kassim, M. R. M., Harun, A. N., & Yusoff, I. M. (2018, November). Smart agriculture using internet of things. In: *2018 IEEE Conference on Open Systems (ICOS)* (pp. 54–59). IEEE, Langkawi.

Meijer, G., Makinwa, K., & Pertijs, M. (Eds.). (2014). *Smart Sensor Systems: Emerging Technologies and Applications*. Wiley.

Muangprathub, J., Boonnam, N., Kajornkasirat, S., Lekbangpong, N., Wanichsombat, A., & Nillaor, P. (2019). IoT and agriculture data analysis for smart farm. *Computers and Electronics in Agriculture*, *156*, 467–474.

Nayyar, A., & Singh, R. (2015). A comprehensive review of simulation tools for wireless sensor networks (WSNs). *Journal of Wireless Networking and Communications*, *5*(1), 19–47.

Pan, M. S., & Tseng, Y. C. (2007). ZigBee and their applications. In: M. S. Pan & Y. C. Tseng (eds.), *Sensor Networks and Configuration: Fundamentals, Standards, Platforms, and Applications* (pp. 349–368). Springer.

Patil, P. A., Jagyasi, B. G., Raval, J., Warke, N., & Vaidya, P. P. (2015, January). Design and development of wearable sensor textile for precision agriculture. In: *2015 7th International Conference on Communication Systems and Networks (COMSNETS)* (pp. 1–6). IEEE.

Patil, V. C., Al-Gaadi, K. A., Biradar, D. P., & Rangaswamy, M. (2012). IoT and cloud computing for agriculture: an overview. In: *Proceedings of Agro-Informatics and Precision Agriculture (AIPA 2012)*, India (pp. 292–296).

Preti, M., Verheggen, F., & Angeli, S. (2021). Insect pest monitoring with camera-equipped traps: strengths and limitations. *Journal of Pest Science*, *94*(2), 203–217.

Qu, C. C., Sun, X. Y., Sun, W. X., Cao, L. X., Wang, X. Q., & He, Z. Z. (2021). Flexible wearables for plants. *Small*, *17*(50), 2104482.

Rajak, P., Ganguly, A., Adhikary, S., & Bhattacharya, S. (2023). Internet of Things and smart sensors in agriculture: scopes and challenges. *Journal of Agriculture and Food Research*, *14*, 100776.

Rajaraman, V. (2017). Radio frequency identification. *Resonance*, *22*, 549–575.

Ramírez-Moreno, M. A., Keshtkar, S., Padilla-Reyes, D. A., Ramos-López, E., García-Martínez, M., Hernández-Luna, M. C., ... & Lozoya-Santos, J. D. J. (2021). Sensors for sustainable smart cities: a review. *Applied Sciences*, *11*(17), 8198.

Ramya, C. M., Shanmugaraj, M., & Prabakaran, R. (2011, April). Study on ZigBee technology. In: *2011 3rd International Conference on Electronics Computer Technology* (Vol. 6, pp. 297–301). IEEE, Kanyakumari.

Rivera, G., Porras, R., Florencia, R., & Sánchez-Solís, J. P. (2023). LiDAR applications in precision agriculture for cultivating crops: a review of recent advances. *Computers and Electronics in Agriculture*, *207*, 107737.

Roberts, C. M. (2006). Radio frequency identification (RFID). *Computers & Security*, *25*(1), 18–26.

Saifuzzaman, M., Khan, A. H., Moon, N. N., & Nur, F. N. (2017). Smart Security for an Organization based on IoT. *International Journal of Computer Applications*, *165*(10), 33–38.

Schütze, A., Helwig, N., & Schneider, T. (2018). Sensors 4.0–smart sensors and measurement technology enable Industry 4.0. *Journal of Sensors and Sensor Systems*, *7*(1), 359–371.

Shaked, B., Amore, A., Ioannou, C., Valdés, F., Alorda, B., Papanastasiou, S., ... & Nestel, D. (2018). Electronic traps for detection and population monitoring of adult fruit flies (Diptera: Tephritidae). *Journal of Applied Entomology*, *142* (1–2), 43–51.

Sharma, P., & Prakash, S. (2021). Real time weather monitoring system using IoT. *ITM Web of Conferences*, *40*, 01006.

Shenoy, J., & Pingle, Y. (2016, March). IOT in agriculture. In: *2016 3rd International Conference on Computing for Sustainable Global Development (INDIACom)* (pp. 1456–1458). IEEE.

Sinha, B. B., & Dhanalakshmi, R. (2022). Recent advancements and challenges of Internet of Things in smart agriculture: a survey. *Future Generation Computer Systems*, *126*, 169–184.

Subashini, M. M., Das, S., Heble, S., Raj, U., & Karthik, R. (2018). Internet of things based wireless plant sensor for smart farming. *Indonesian Journal of Electrical Engineering and Computer Science*, *10*(2), 456–468.

Tiwari, M. M., Narang, D., Goel, P., Gadhwal, A., Gupta, A., & Chawla, A. (2020). Weather monitoring system using IoT and cloud computing. *Weather*, *29*(12s), 2473–2479.

Tiwari, V. K., & Singh, V. (2016). Study of IoT: a vision, architectural elements, and future directions. *International Journal of Advanced Research in Computer Science*, *7*(7) 65–84.

Ullo, S. L., & Sinha, G. R. (2020). Advances in smart environment monitoring systems using IoT and sensors. *Sensors*, *20*(11), 3113.

Ullo, S. L., & Sinha, G. R. (2021). Advances in IoT and smart sensors for remote sensing and agriculture applications. *Remote Sensing*, *13*(13), 2585.

Vongsingthong, S., & Smanchat, S. (2014). Internet of things: a review of applications and technologies. *Suranaree Journal of Science and Technology*, *21*(4), 359–374.1

Wang, Q., & Balasingham, I. (2010). Wireless sensor networks-an introduction. In: T. Y. Kheng (ed.), *Wireless Sensor Networks: Application-Centric Design* (pp. 1–14). IntechOpen.

Xia, M., & Song, D. (2018). Application of wireless sensor network in smart buildings. In: *Machine Learning and Intelligent Communications: Second International Conference, MLICOM 2017,* Weihai, China, August 5–6, 2017, *Proceedings, Part I 2* (pp. 315–325). Springer.

Xiao, Y., Yu, S., Wu, K., Ni, Q., Janecek, C., & Nordstad, J. (2007). Radio frequency identification: technologies, applications, and research issues. *Wireless Communications and Mobile Computing*, 7(4), 457–472.
Yassein, M. B., Mardini, W., & Khalil, A. (2016, September). Smart homes automation using Z-wave protocol. In: *2016 International Conference on Engineering & MIS (ICEMIS)* (pp. 1–6). IEEE, Agadir.
Yin, H., Cao, Y., Marelli, B., Zeng, X., Mason, A. J., & Cao, C. (2021). Soil sensors and plant wearables for smart and precision agriculture. *Advanced Materials*, *33*(20), 2007764.

Chapter 3

Utilization of smart sensors in localization, navigation, and mapping

Sanjeevi Ramakrishnan, Anuradha Jayaraman, Sandeep Tripathi, and Prashantkumar B. Sathvara

3.1 INTRODUCTION

In recent days, the proliferation of unmanned aerial vehicles (UAVs), smart cities and vehicles, and the Internet of Things (IoT) has underscored the critical role of navigation systems. From inertial navigation to the array of satellite-based systems like the global navigation satellite system (GNSS) comprising BDS, GPS, GLONASS, and Galileo, alongside emerging technologies such as Bluetooth, Wi-Fi, ultrasonic, Ultra Wideband (UWB), radio frequency identification (RFID), magnetic, odometer, light detection and ranging (LiDAR), vision, and 5G, a wide spectrum of cutting-edge navigation systems has emerged. These systems incorporate diverse sensor types like magnetometers, photodiodes, wireless receivers, GNSS receivers, inertial sensors, and cameras. However, relying solely on single-sensor-based navigation systems poses significant challenges in delivering a dependable, precise, and seamless navigation solution [1]. For this case, the inertial navigation system (INS) provides precise comparative positioning for a limited duration until it begins to diverge due to inherent inertial sensor errors and integration issues. Thus, to maintain accuracy over time, supplementary absolute positional data sources such as GNSS are commonly integrated. However, GNSS encounters challenges in urban environments and other obstructed areas, leading to signal blockage and multipath interference. Additionally, alternative wireless location systems like Wi-Fi and Bluetooth present their drawbacks, including reliance on access point (AP) distribution, inconsistent and noisy solutions, resource-intensive database construction, and variable received signal strengths (RSS) indoors [2]. In indoor environments, local positioning methods such as magnetic positioning are commonly employed instead of global systems. Vision-based positioning, utilizing cameras to track object movement, proposes precise localization at a reasonable charge. However, it comes with drawbacks like privacy implications, the necessity to interpret surroundings for data extraction, and a significant processing burden. Consequently, achieving application-specific requirements and navigation objectives solely through a single-sensor approach presents considerable challenges.

DOI: 10.1201/9781003633884-3

Consequently, leveraging the strengths of different sensor types enables improved positioning performance, enhancing robustness, reliability, and coverage across time and geography. Amongst the integrated utilization of GNSS/INS system, GNSS aids in minimizing cumulative errors by offering position and velocity data to support inertial navigation. Meanwhile, INS complements GNSS by filling in gaps in challenging environments and filtering its inherent noise. Thus, integration mitigates the limitations of each component, leading to enhanced system performance [3].

Various positioning systems such as Bluetooth, Wi-Fi, Vision, UWB, magnetic, RFID, and ultrasonic can be integrated with INS. These integrations are typically categorized into three main types: (1) tightly coupled integration (TCI), (2) loosely coupled integration (LCI), and (3) ultra-tightly coupled integration (UTCI). TCI involves integrating sensor data within the range levels between transmitters and the receiver. Conversely, LCI integrates multiple sources based on position and velocity levels [4]. UTCI encompasses raw metrics from a GNSS receiver at a more profound level compared to LCI and TCI, incorporating data such as code and raw carrier phase. Multi-sensor integrated systems find applications in various domains, including automobiles, spacecraft, surface and underwater vessels, aircraft, indoor mobile robots, and UAVs.

The cornerstone of integrated systems lies in dataset fusion, which amalgamates diverse evidence from multiple studies. The Joint Directors of Laboratories (JDL) workshop established a widely accepted definition of data fusion, stating that it involves a multilevel procedure dealing with the analysis of data, combination, correlation, and association, and association information from various sources to obtain more precise positions, identify estimates, and provide comprehensive and timely evaluations of events and threats. While various techniques exist, analytics- and learning-based methods are among the most prevalent approaches [5]. In analytics-based approaches, system states and external measurements, often termed "estimations" in literature, are represented using analytical functions. Conversely, numerous learning-based data fusion techniques, such as artificial neural network (ANN), fuzzy logic, and support vector machine (SVM), are employed to process and integrate data [6]. Learning-based techniques are favored due to their ability to model systems without prior statistical information regarding process and measurement noise. However, analytics-based methods, such as the particle filter, remain widely utilized. In certain instances, learning-based approaches demonstrate superior performance compared to analytics-based methods.

We surveyed existing literature concerning data fusion for integrated navigation systems to improve the navigational capabilities of autonomous underwater vehicles (AUVs) [7]. We examined the progressions in multi-sensor data fusion techniques and their applications in AUV navigation. This exploration included the implementation of various algorithms across different layers of the JDL model [8]. They also highlighted both the limitations and benefits of such applications. Additionally, a classification of vision-based fusion strategies in unstructured environments was provided [9]. In their assessment of wireless-based indoor location solutions, Yassin and Elsadig [10] essentially introduced hybrid systems that combine cameras, inertial sensors, and map matching. They also introduced fusion-based indoor positioning methods and systems characterized by three fusion aspects: source, algorithm, and weight spaces [11].

The framework was the main topic of the study; navigation sources and application scenarios were not included. Additionally, both are outdated. The poll made no mention of learning-based approaches and exclusively addressed vision-based positioning [12]. The essay offered a limited range of indoor sensor fusion, but it explored wireless positioning methodically. The study conducted a thorough examination of integration sources, encompassing both analytics-based and learning-based methodologies. However, they only cover indoor placement, and their talks of algorithms and design consideration are scant. Furthermore, neither of these studies specifically addressed integration scheme-based classifications or optimization-based techniques.

Three levels are used to organize the text: application scenarios, integration techniques and architectures, and sensors. Figure 3.1 shows the system's general design. The primary inputs are as follows:

- Multi-sensor fusion is categorized into three types: relative/relative, absolute/absolute, and absolute/relative integration based on sensor integration. Absolute positioning sources are further subdivided into five groups based on their physical characteristics: audio, radio, field, light, and vision-based. The study extends its scope to explore various multisensory combinations.
- The primary focus lies between learning and analytics-based systems leveraging database fusion methods. These classifications are often developed across multiple disciplines and scenarios. Specialized techniques like visual-inertial odometry (VIO) are not extensively covered. The survey of analytics-based systems' algorithms and integration architectures considers three perspectives: ultra-tightly coupled, tightly coupled, and loosely coupled methods.
- Integration designs and algorithms based on analytics are discussed, highlighting the evolution from filtering-based navigation systems to more versatile analytics-based algorithms applicable to graph optimization and filtering techniques. Learning-based navigation and location techniques are categorized as deep, reinforcement, unsupervised, and supervised learning.
- Application situations are separated into six groups, with examples and analyses provided for each. Specific themes such as autonomous vehicles and UAVs are explored.

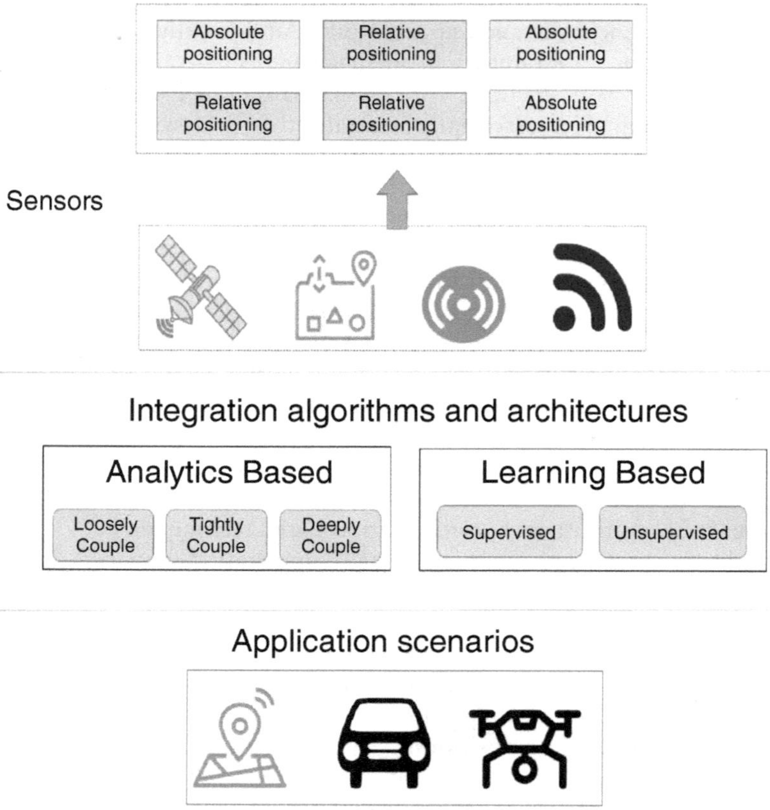

Figure 3.1 The comprehensive structure of the entire navigation system with application scenarios.

- A comprehensive guide on design considerations is offered, covering aspects like choosing between learning and analytics-based structures, selection of state and observation, real-time considerations, synchronization of time, scalability, and outlier resilience. Future research directions in integrated navigation are also discussed.

The study covers INSs, diverse wireless location structure systems, and multisource systems. It presents multi-sensor analytical-based big data fusion techniques and discusses design considerations and future research paths in integrated navigation.

3.2 POSITION OR NAVIGATION SOURCES

In this part, we introduce popular standalone navigation systems and categorize combined navigation systems based on source types.

3.2.1 Single navigation systems

The commonly used single navigation systems such as GNSS, Bluetooth, Wi-Fi, RFID, odometer, magnetics, vision, LiDAR, and 5G are presented here [13]. Certain upcoming positioning technologies are not addressed since they have not yet been implemented, such as terahertz-band localization for 6G.

3.2.1.1 Global navigation satellite system

To offer real-time, all-land, air navigation, weather condition location, and sea navigation services for intelligence collection, emergency communications, and nuclear explosion monitoring, the GNSS was technologically advanced in the 1960s. A position may be estimated using GNSS to a precision of a few meters to centimeters by using the arrival of time in combination with information from many communication navigation satellites. However, because of signal blockage, multipath, and electromagnetic interference, GNSS is imprecise inside and in urban canyons [5]. To overcome these limitations, it is essential to use in conjunction with additional local positioning technologies such as UWB, INS, and Wi-Fi. INS is an ideal supplemental data source since integration reduces the disadvantages of inertial dead-reckoning and the GNSS's line of sight [14]. If there is a disruption or obstruction in the GNSS signal, the inertial sensor detection permits the system to coast until the signal is recovered. The combination of GNSS and INS improves the integrity and quality of each component and permits the correction of inertial instrument biases, while the inertial sensors advance in the performance of the GNSS receiver.

3.2.1.2 Inertial navigation

Several decades ago, inertial navigation was created employing inertial sensors, like gyroscopes and accelerometers for use in military hardware, such as ships, aircraft, and missiles. There are several grades of inertial sensors, including low-cost, auto, tactical, and navigation; the differences in their cost and performance are notable. Through methods like using information from accelerometers and gyroscopes, inertial navigation (INS mechanization, motion limits, and pedestrian dead reckoning, or PDR) determines the platform's velocity position, and orientation. The accuracy of inertial navigation algorithms decreases accumulatively, contingent on inertial sensor faults, due to errors in integral computations and inertial sensors. Since inertial navigation is based on self-contained inertial sensors, it functions without external data and is unaffected by electromagnetic interference from the outside world [15]. Inertial navigation applies to both cars and pedestrians and may be employed in a variety of situations, including the sky, ground, and underwater. With a high update rate, strong stability, and location, velocity, and attitude solutions, it has the disadvantage of low long-term accuracy due to increasing navigational mistakes over time.

3.2.1.3 Wi-Fi

With compatibility with many IEEE 802.11 protocols, Wi-Fi is currently utilized virtually everywhere, from IoT to smart cities. Its hardware typically consists of Wi-Fi routers, also known as APs, acting as transmitters and Wi-Fi receivers acting as receivers. Wi-Fi has been studied for around 20 years. Common measures in Wi-Fi location systems include flight durations, channel state information (CSI), arrival angle, and RSS. They frequently employ a variety of algorithms, including fingerprinting and trilateration. The benefits of Wi-Fi-based positioning systems include wide coverage, low cost, and the ability to support smart devices. They also offer adequate positioning precision (between 1 and 10 m) and decent mobility. Furthermore, they do not require extra infrastructure. Nevertheless, multipath, signal obstructions, AP dispersion, measurement fluctuations, and high labor expenses associated with database construction can all interfere with Wi-Fi transmissions. Analytics-based methods, such as optimization strategies, utilization in the Wi-Fi or inertial system for insider robot and pedestrian applications algorithm in terms of data fusion algorithms.

3.2.1.4 Bluetooth

Bluetooth is a data exchanging system for short-range detection that uses shorter wavelength ultra-high-frequency radio waves in an industrial-based scientific medical band with 2.400–2.485 GHz frequency. Because of its low power consumption, it is frequently utilized in IoT applications and is governed by the Bluetooth Special Interest Group. Similar to Wi-Fi positioning, the Bluetooth Low Energy (BLE) system often makes use of beacons for both transmitter and receiver functions, whereas RSS data are used for positioning. Proximity, multilateration, fingerprinting, and the combination of multilateration and fingerprinting are the categories of algorithms. The benefits of BLE beacons are low weight, cheap cost, low dissipation, compact size, and widespread compatibility with smart devices. BLE uses less energy, supports more smart devices, is more versatile, and is simpler to implement than Wi-Fi [16]. Nevertheless, it shares many of the same drawbacks as Wi-Fi, including high labor costs associated with creating fingerprint databases and performance degradations brought on by multipath, signal obstructions, and RSS volatility. Moreover, BLE devices and Wi-Fi typically interfere with BLE systems since they share the same

2.4 GHz band. According to the literature, RSS and occasionally time of arrival (TOA) are responsible for implementing BLE/INS integrated navigation. Furthermore, the precision of its angle of arrival (AOA) location has increased since the release of Bluetooth 5.1 protocols in 2019.

3.2.1.5 *Ultra-wideband*

UWB wireless communications function in three categories: location, radar, and communications. It does not require a carrier and communicates in a very short time using a pulse. Its positioning algorithms are divided into groups: RSS, AOA, TDOA, and TOA. It is effective in non-line-of-sight scenarios thanks to its large bandwidth and very short pulse waveforms. Nevertheless, additional infrastructure is required, and liquids and metals can interfere with signals. Misconfigured devices that operate in the ultra-wideband can also cause interference. Also, a longer synchronization period might result from the brief pulses.

3.2.1.6 *Visible light positioning*

Using optical light instead of radio waves, visible light communication is a relatively new method for data transmission. Compared to current systems, it has several advantages such as high data rates, extended lifespans, low heating, and excellent energy efficiency, all without posing a risk to human health. Accordingly, it is currently used for location in the navigation sector and presents a new approach positioning method called VLP. In a true VLP structure, the transmitters are light-emitting diode (LED) bulbs, and the receiver is a photodiode, also called a camera. The analysis of optical information may be done using a variety of modalities, such as RSS, TOA, TDOA, and AOA [17]. For each mode, many location techniques such as proximity, trilateration, multilateration, and fingerprinting have been applied in recent studies. Most VLP systems only achieve precision of a less centimeters since the optical signal is immune to electromagnetic wave interference, high multipath interference, and fading. However, shadowing or blocking impacts VLP because optical signals cannot pass through opaque surfaces.

3.2.1.7 *Radiofrequency identification*

A popular technology in many IoT applications is RFID. To compute the separation between the receiver and transmitter, it makes use of AOA and radio frequency (RF) RSS. The accuracy of its placement ranges from centimeters to rooms, depending on how tags are distributed and which positioning algorithms are selected. You may choose where to put its tags, which are semi-active, passive, and active. Some of the enticing features that have added to its attractiveness include remote communications, quick record speeds, non-line-of-sight reading, strong security, less amount, and compactness. Using this placement strategy, an RFID card reader is attached to a monitored object after passive and active tags have been located on the floor and ceiling. Two potential application situations are inside-location pedestrian navigation and positioning.

3.2.1.8 Ultrasonic

Inspired by bats' ability to navigate at night, ultrasonic systems have been researched extensively and categorized into two categories: (1) using a tag attached to the item that serves as a receiver and many transmitters affixed to a wall or ceiling [8], and (2) by mounting a tag on the item that serves as a transmitter and placing several receivers on a wall or ceiling. Trilateration and proximity are the primary methods used in ultrasonic positioning to process TOA or TDOA signals, depending on the level of precision required and the operator's financial constraints [11,18]. When there are many closely spaced nodes, ultrasonic can reach centimeter-level precision; when there are fewer widely spaced nodes, it can reach room-level accuracy. Numerous benefits come with this system, but the main one is that there is no leakage between rooms. Other benefits include cheap system cost, great dependability and scalability, and high energy efficiency. However, because humidity and temperature may readily alter the sound's transmission speed in the air, its performance may fluctuate depending on the surroundings. Moreover, ambient noise and reflected ultrasonic waves might impair its performance.

3.2.1.9 Odometer

An odometer is a device that measures a land vehicle's wheel rotation and provides data on the speed and distance traveled. Traditionally, odometers were mounted on the gearbox shaft, but the majority of modern cars have one on each wheel. Thus, a wheel speed sensor is another name for an odometer. A differential odometer is the process of measuring the vehicle's yaw rate by separating the left and right odometer readings. It should be noted that the term "odometer" should not be confused with "odometer," which describes a method that uses motion sensor information (such as LiDAR, odometers, and cameras) to estimate position change over time. Odometers have been used in numerous research on the integration of GNSS and INS since they are often encountered on ground vehicles [19,20]. During extended GNSS outages, it is helpful to increase position, velocity, and attitude accuracy.

3.2.1.10 Magnetics

Most very low-frequency static magnetic field applications in the short area range orientation and position measurements are based on open free-space field area geometry. They provide a locating resolution with many meters of accuracy without the help of other systems, and they can counter some wireless location technologies. Artificial indoor disturbances brought on by electrical currents in metal or other conducting items may interfere with wireless-based navigation [21,22]; however, the magnetic anomalies these interferences produce can be utilized as landmarks or fingerprints. Magnetic solutions can lead to large localization errors, so they are not always appropriate.

3.2.1.11 Vision

Wireless-based systems lack the robustness of vision-based systems, which make use of smart gadgets and security cameras to determine the position of objects through photos or videos. They fall into two categories: (1) those with cameras mounted at fixed locations and (2) those with cameras mounted on mobile machines. In the category of image-based camera position, two distinctions emerge: those operating within a known environment and those navigating unknown surroundings. In a familiar setting, diverse types of information such as points, geometric data, and semantic cues can be leveraged to estimate the absolute perfect pose (position+orientation) to the camera. This challenge is often referred to as the Perspective-n-Point problem. These methods often rely on prior data from other sensors, such as GNSS and magnetics, especially when deployed in outdoor environments [16]. That is a structure from motion (post-time) or simultaneous position and mapping (real-time) challenge in an unfamiliar environment condition. Visual odometer (VO) and VIO, which exclusively solve relative pose, are the two common simultaneous localization and mapping (SLAM) positioning methods. Visual tracking, another name for the second group, is capable of being combined with motion detection [20,23]. Pedestrian Dead Reckoning (PDR) with visual tracking using an installed camera that watched a lengthy corridor is proposed in [24].

It has been extensively studied how to map and measure visual odometers. As a potential substitute for traditional odometers, a vision odometer (VO) uses a camera mounted on the agent to gather picture data and associates homologous spots in several photos to estimate an agent's position. A few popular vision-based algorithms include OV2SLAM, DSO, ORB-SLAM, and MonoSLAM [2,25]. Readers may consult for an overview of monocular SLAM, as well as for an overview of mapping and visual odometer techniques that use deep learning as their main tool [26].

The combination of vision and the inertial measurement unit (IMU) for motion estimation opens up a wide range of robotics applications. Because visual and inertial sensors complement one another, a legged or wheeled robot functioning in an indoor, outdoor, or industrial setting serves as an excellent example of VIO's performance as an inertial navigator [27]. There is a wealth of research on visual-inertial odometer and mapping; noteworthy studies include ORBSLAM3, OKVIS, VINS-Mono, KSWF, MSCKF, and OpenVINS. For a survey of visual-inertial odometers, as well as a more current survey based on deep learning, detailed in Zhuang et al (2023).

3.2.1.12 5G networks

5G (5th Generation Mobile Communication Technology) position takes advantage of cellular networks. Whereas the localization accuracy of 2G, 3G, and 4G was restricted, 5G can achieve range accuracy down to the centimeter. Additionally, greater denser networks, signal bandwidths, time of flight (TOF), and multiple-input multiple-output technology positioning strategies all significantly increase range accuracy. Among the frequencies allocated for 5G are mm-wave bands and sub-6 GHz, both of which can be used for localization. Studies on 5G positioning have mostly employed simulated data rather than real testing, as 5G has only recently been put into use. Utilizing a 32-dual-polarized base station antenna array, the authors of Ref. [28] collected CSI data and utilized extreme learning machines to predict the location of a single antenna [29]. Pseudo-range measurements on 600 MHz bands using a software-defined receiver; 1.19 m was their ranging error. In field tests conducted indoors [13], TOA ranging accuracy of 0.5 m was recently achieved using 5G new radio carrier-phase measurements in the sub-6-GHz region.

3.2.1.13 LiDAR

An essential part of self-governing vehicles, like UAVs, and self-driving cars is the 3D LiDAR sensor. It can measure the depth of its surroundings directly, densely, actively, and precisely. Using a 10-Hz high sampling rate [12], LiDAR gathers point cloud data of nearby objects, making it beneficial for figuring out how the vehicle's position has changed. LiDAR odometer, another term for LiDAR placement, is primarily a relative positioning technique [12]. The most traditional real-time LiDAR odometer is Lidar Odometry and Mapping (LOAM) [22], which builds a global map progressively by successively registering extracted edge and planar features. Additional research has been conducted using LOAM, such as LeGO-LOAM and Fast-LOAM.

LiDAR-inertial odometer (LIO), which compensates for gesture distortion in a LiDAR scan while offering a reliable beginning point, may significantly improve the accuracy and resilience of LiDAR odometer with the aid of inertial sensors. To improve the LiDAR posture in the odometer, LINS included a closely connected iterative robot-centric formula. A novel data structure called ikd-Tree was introduced by FAST-LIO2 to facilitate fast queries and incremental map changes at each step [30]. When LIO-SAM originally developed the LIO odometer, several relative and absolute metrics, including loop closure, might be added to the system as factors from different sources as a factor graph. Self-driving systems can use GNSS, IMU, LiDAR, cameras, and radar to perform a range of functions.

3.2.2 Navigation systems with several sensors integrated

Section 3.1 gives definitions for absolute and relative positioning. Three categories—absolute/absolute integration systems, absolute/relative integration, and relative/relative integration—were used in this area to group multisensory integrated systems. Based on this definition, this categorization was made.

3.2.2.1 Absolute positioning sources

Absolute positioning is the process of determining an object's attitude and location in a predetermined position frame. It can be classified as audio, radio, field, light, and vision-based, in that order. Radio-based location positioning comes in a variety of forms, including Bluetooth, GNSS, RFID, Wi-Fi, 5G, and UWB. They use artificial production sources to generate electromagnetic waves for localization in RF bands. The most popular kind of light-based positioning is called VLP (Visible Light Positioning), and it makes use of photodiodes, or cameras, as the receiver and LED lights as transmitters. Most applications of audio-based location systems take place inside, and they include acoustic and ultrasonic in the audible spectrum [31]. With INS, location by acoustic and ultrasonic means may be combined. For further information on other kinds of audio-based hybrid systems, by matching the database holding field data, field-based placement navigates. Field-based positioning sources that are frequently employed are maps, magnetics, and gravity. Gravity is corresponding can be used in the downside of water passive range navigation systems to get around the INS error gathering and on landscape areas with the terrain shape matching. Map identical, which is also available in city settings, can assist additional sensors with reliable location limitations in indoor positioning. Magnetic-based positioning is primarily used indoors. With vision, one can achieve both exact and relative positioning. Furthermore, synthetic aperture radar (SAR), a form of generalized visual location technology, employs active microwave imaging radar.

3.2.2.2 Relative positioning sources

Relative sources of positioning, which identify an object's location and position in a movement standard frame or simply monitor its rotational and velocity rate, primarily use inertial sensors and eyesight. Since vision and INS have already been thoroughly covered, we focus more on pedestrian dead reckoning (PDR), another inertial system, in this section. PDR systems detect user movements by utilizing inertial sensors in conjunction with walking domain-specific information. PDR calculates a pedestrian's two-dimensional (2D) location by adding up (distance, heading) vectors, which can be either steps or strides. Cheap Micro-Electro-Mechanical Systems (MEMS) sensors become commonplace as smartphones gain traction in the market, which leads to the widespread use of PDR. Radio-based location technologies found on smartphones, such as Wi-Fi, BLE, and UWB, can be linked with PDR to stop error accumulation.

3.2.2.3 Absolute/relative integration

The most common form of integrated navigation system combines one absolute positioning system with one relative system for positioning. Since most mixture systems have inertial sensors, we divide this form of integration into groups: Field, field/inertial, radio, light, audio, and vision-based are the five categories of inertial fusion. These inertial systems include the PDR and INS. It is quite uncommon to find an integration system, such as GPS/VO, that has relative positioning but lacks inertial systems.

Because these five groups are physically distinct from one another, their applications and benefits also vary. The most common integrated navigation system is radio/inertial due to the extensive use of GNSS and Wi-Fi, as well as the widespread use of BLE. While this category's area of coverage and system capacity can be quite substantial, multipath interference and Non-Line of Sight (NLos), particularly indoors, might affect its accuracy. Light-based positioning employs light waves to position, much like radio does. While VLP is more precise indoors, it is more susceptible to coverage restrictions and signal obstructions. Additionally, it is less susceptible to electromagnetic wave interference, multipath, and fading. Compared to electromagnetic waves, audio waves propagate at a substantially slower speed. As such, measuring the time of flight (ToF) is considerably simpler [30,32]. While precise in range, audio-based location is not reliable in dynamic situations. Field-based placement, as opposed to the other three categories, is unaffected by outside disturbance. On the other hand, complex sceneries could include many localization components (such as a positioning system for a capsule endoscope), which is why field area matching is a perfect method to reward other absolute location and positioning systems. Vision fusion-based systems do not incorporate VIO and VO since, in this context, vision refers to complete position. GNSS, magnetics, BLE, Wi-Fi, and inertial sensors are some of its fusion sources.

3.2.2.4 Relative integration

Two relative positioning devices are among the fusion systems, in addition to the usual absolute/relative integration. The majority of these systems use inertial sensors. VIO is a popular option to help INS delay the error buildup due to the development of computer vision and the inexpensive cost of cameras [33]. Visual-inertial integration may be used in many different scenarios, such as space exploration, surface ships, interior and outdoor vehicles, and more. An absolute location source can help visual-inertial systems to further limit error buildup.

Depending on the application circumstances, other relative positioning technologies such as LiDAR, PDR, Doppler velocity log (DVL), speed log, odometer, and speed log can assist INS. All of the instruments that measure speed are the DVL and the odometer. Since land vehicles often have odometers, it makes perfect sense to incorporate inertial sensors with them. Speed logs can be used as a substitute for odometers by ships and AUVs. They use the Doppler effect to calculate velocities. Even though they could have the same inertial sensors, pedestrian users can interface with INS via PDR support. LiDAR's miniaturization opens up new possibilities for land vehicles, particularly autonomous driving.

3.2.2.5 Absolute/absolute integration

There are additional works that use two sources of absolute location without relative positioning. Despite having comparable physical characteristics, it is possible to balance out two radio-based systems by combining them. GNSS/5G, Wi-Fi/BLE, plus RFID/wireless sensing network (WSN) are all involved in these projects. Radio-ranging-based geolocation technologies, such as dedicated immediate connection and UWB, can compute vehicle-to-vehicle distances for cooperative localization since radios may connect cars via channels for communication. Nevertheless, these systems may have low sample rates and offer less orientation and velocity data than inertial sensors.

Using an inertial sensor to merge two precise location sources is one way to use INS-free fusion systems. In protected settings, GNSS is not very good at providing accurate locating results; however, alternative radio-based sources can, given the necessary infrastructure. To further enhance performance in GNSS-challenging scenarios, Wi-Fi, RFID, BLE, and UWB can be coupled with the GNSS/INS system in this manner. Similarly, two radio-based systems (such as Wi-Fi/BLE/INS and RFID/UWB/PDR integration) can be combined with the inertial sensors, except GNSS.

Since every absolute positioning category has unique physical characteristics, multi-category fusion might be a more effective way to balance each other out. The two most commonly used positioning sources in the majority of fusion systems having two categories for absolute positioning are field-based and radio-based. Since radio-based arrangements are vulnerable to signal obstruction while field systems are not, they cancel each other out. A few of the works include Wi-Fi/map, UWB/magnetics, GPS/map, and GPS/magnetics, choose radio/field integration. Since RF bands diverge significantly from light, radio-based location-assisted variable-length

Table 3.1 Artificial intelligence-based model algorithms used in the navigation system

S. no.	*Algorithms*	*Usages*	*References*
1	Artificial neural network	Forecast locations during GNSS interruptions Complete positioning solution	[35]
2	Fuzzy logic	Modify covariance matrices for measurement and process noise	[36]
3	Support vector machine	Forecast within fusion systems in the event of sensor outages	[37]
4	Random forest	Identify human activity Assess the accuracy of GNSS positioning Model significant nonlinear errors in INS	[38]
5	Deep learning	Complete positioning solution Handle camera and laser data processing	[39,40]
6	Reinforcement learning	Adjust the navigation parameters Vehicle tracking	[37,39,41]

pulses are useful with reference to radio/vision fusion [34], an INS/GPS/SAR system in which SAR gives azimuth and pitch while the combination of Global Positioning System (GPS) and INS provides pseudo distances and their change rates. Furthermore, the works integrated magnetics and ultrasonic indoors, with magnetics providing attitudes and ultrasonic providing distances.

3.3 MULTI-SENSOR DATA GATHERING FOR LEARNING-BASED INTEGRATED NAVIGATION

An overall classification of the learning-based techniques utilized in integrated navigational systems is provided in this section. They can be roughly classified into two groups: end-to-end placement and assisting parameter estimate in analytical procedures. It is noteworthy that the majority of innovative contributions rely on combining multiple learning-based algorithms (Table 3.1).

3.4 APPLICATION SCENARIOS

3.4.1 Land vehicles position

Land vehicles have high localization and navigation requirements. Examples of these vehicles include automobiles, motorbikes, vans, and mobile robots. We choose mobile and automobile robots as our representations in this area. The most popular navigation application, car navigation, has drawn the attention of several academics. Presently, the most widely used positioning system is GNSS/INS integrated navigation. An accurate and reliable navigation system may be produced using GNSS with the help of odometers, vision, LiDAR, ultrasonics, map matching, and UWB in some settings (such as metropolitan areas) where it suffers greatly from multipath. LiDAR and vision have gained popularity recently in automotive navigation.

Industry investment in autonomous driving has surged recently. Unlike automotive navigation systems, it does perception, course planning, control, and localization in addition to these other functions, where localization is crucial. Based on the interdependency of the following, there are three main approaches to joining sensory data: high-level, mid-level, and minimal fusion. This is a result of the multitasking nature of autonomous driving. Autonomous vehicles usually combine information from camera-radar-LiDAR, camera-LiDAR, and camera radar using GNSS and IMU. The camera, GNSS, LiDAR, IMU, and millimeter-wave radar are the main instruments used by these vehicles. The study used a pre-made map for LiDAR. The practice of localization and multi-sensor data helped resolve GNSS RTK ambiguity. In congested metropolitan streets, the technology-enabled complete autonomy for cars. To optimize the poses the authors of Ref. [3] incorporated the LiDAR scan matching, visual odometer, and other methods with the GNSS/INS solution. A 2D probability map was then created using the posture that resulted and the measurements from the LiDAR scan. Additionally, following other vehicles is required for several duties (such as path planning, intent identification, and obstacle avoidance) and is a prerequisite for fully autonomous driving. Because LiDAR and cameras have complementary skills for seeing the environment, research has employed them to track several objects. In automated driving, a summary of multi-sensor fusion was released.

The basic criteria for developing an autonomous mobile robot system are localization and navigation. While GNSS, vision-based, and other LiDAR-based systems are frequently used in outside situations, RF-based, and acoustic technologies are popular indoors. It is well known that mobile robot localization can make use of GNSS, RF-based, and auditory approaches combined. Sonar outperforms vision sensors whenever the robot enters a dark region. The combination of internet connections, inertial measurements, and ultrasonic vibrations allowed an indoor robot to locate itself [4]. Additionally, RF signals like Wi-Fi, WSN, and UWB can be added in an interior setting to the localized system.

Over the past ten years, there has been a tremendous advancement in SLAM, opening up new possibilities for camera- and LiDAR-based mobile robot navigation. For robotic navigation applications, several LIO and VIO systems have been developed, with FAST-LIO2 and VINS-Mono. EKF combined vision-based navigation systems with an inexpensive, reduced inertial sensing system that uses IMUs and wheel speed sensors to fill in the navigation gaps during GPS outages. For mobile robots, [26] introduced Integrated Circuit-based Global Navigation Satellite System and Visual Inertial Navigation System (IC-GVINS), which completely utilizes the accurate INS in the process of visualizing and estimating state.

3.4.2 Indoor position localization

The vision-based, momentum, RF, ultrasound, and magnetic positioning systems are the five main systems used for indoor localization. However, a unified approach is required indoors because a single geolocation method is unreliable. Typically, a mix of inertial and wireless sensors is employed. Estimation filters were included in Wi-Fi/PDR navigation systems to improve them. It is also possible to merge Wi-Fi signals with INS. Although most buildings have Wi-Fi APs, localization systems do not require further infrastructure because of their high power consumption [42]. Since Bluetooth uses less power than Wi-Fi, the idea of combining Bluetooth with inertial sensors was conceived. Both Wi-Fi and Bluetooth may be used with mobile phones without the need for further hardware, although their precision is poor.

Other configurations are integrated with either PDR or INS via Wireless Local Area Network (WLAN), UWB, and RFID. Systems of this type need more positioning hardware. The integration of visible light positioning into the positioning system is an innovative approach to indoor localization. Positioning precision often exceeds 0.1 m, which is superior to the previously listed technologies. Navigation systems can also benefit from the use of map data and landmarks, even if creating them requires labor and time. Inertial sensors were aided by geomagnetic data.

3.4.3 Air vehicles

Air vehicles, like guided missiles, UAVs, high altitude platform stations, and airplanes, seldom have access to non-GNSS radio-based identifying sources and map data in comparison to land vehicles. In this paragraph, we have designated UAVs and airplanes as the representatives. In multisensory integrated navigation, INS/GNSS addition is the most commonly utilized technology for aircraft and UAVs.

Airborne earth observation now mostly uses an integrated system that combines GNSS and Strapdown Inertial Navigation System (SINS) to extract aircraft trajectory data. The GPS/INS integration was subjected to a sigma-point filter. The covariance matrix was made smaller by representing attitude as a vector of generalized Rodrigues parameters in three dimensions. For INS/GPS integration, an iterated Kalman Filter (KF) was suggested, and it was used for SAR motion correction. Using the Unscented Kalman Filter (UKF) and Cubature Data-driven Kalman Filter (CDKF), the authors of Ref. [43] looked at the alignment of an integrated SINS/GPS while moving. An INS was assisted by an airplane dynamic model to deal with the lack of GPS signals. The application of a direct filtering technique instead of indirect filtering defined by linearized error equations was made possible by a fast-update aviation dynamic model (ADM), nonlinear INS dynamical as system formulas, and a nonlinear ADM as observed equations. A previous study [22] employed a square-root UKF-based data fusion from multiple sensors technique to manage the attitude and position estimations of a small unmanned aircraft attached to a moving platform. Using information from the helicopter's inertial sensors, a kinematic process framework was used to forecast the state. The resulting correction used information from other sensors, including a magnetic, magnetometer, LiDAR altimeter, and barometric altimeter encoders that determined the orientation of the tether concerning the helicopter.

A precise, accurate, and reliable navigation system is necessary for UAVs to perform assigned missions more effectively and save human life, effort, and time. The normal repetitive dynamic patterns of UAVs provided valuable data for the estimation of UAV navigation states. After using machine learning classifiers to identify recurring dynamic patterns, Extended Kalman Filter (EKF) was used to execute the proper constraint or update to improve the estimation of the UAV states. In addition to filters, UAV navigation systems have also made use of smoothers. The location, velocity, and orientation of UAVs were estimated both online and offline. A method known as fixed-interval smoothing was used for offline estimate. Online estimates employed a fixed-lag smoothing method. RF signals, like UWB signals, could be employed in an inside setting to compensate for GNSS signal loss. The authors employed a low-cost IMU and the UbiSense UWB real-time localization system to boost position accuracy [12] to investigate the indoor localization of a small, low-cost UAV. The camera was another sensor that was frequently used in GNSS signal failures. However, low-textured or low-light environments are problematic for vision-based systems. A magnetometer and an IMU were used to provide an indoor UAV with a precise heading estimate.

3.4.4 Surface ships

Since it frequently works in tandem with other sensor paradigms to enhance navigation performance, the GNSS system is an essential component of surface ship navigation systems. The GNSS/INS, GNSS/IMU/speed log, and GNSS/Laser Gyro INS combined systems are a few examples of these combinations. GNSS transmissions, however, have several problems, including spoofs and interceptions. As a result, several researchers have looked at substitutes like radar and the star sensor. Star sensors are state-of-the-art Communication Navigation and Surveillance (CNS) devices that can operate independently to eliminate external interference [44]. A study applied a one-dimensional Wiener filtration method to an integrated SINS/CNS system to achieve recovery of the blurry star image under high dynamical conditions. To circumvent the unreliability of the CNS in overcast situations, the study developed an INS/CNS integrated navigating solution based on optimizing the behavior of particle swarms using back propagation neural network development.

One of the key duties of autonomous ship navigation, which is highly sought after at the moment, is collision avoidance. To do this, radar systems are frequently linked with GNSS or INS. In addition, vision-based navigation is now a topic of intense discussion in the ground of outward ship triangulation since it offers a wealth of information and does not require wireless equipment. A ship's navigation system has measured the distance between a UAV and the ship using an integration of vision, radar, and INS [45]. They created an integrated shipboard-relative GPS/INS/vision system by using a higher sample rate for vision-based navigation. The effort produced positive results by merging INS with ultra-short baseline into a tightly integrated architecture without the use of GNSS. Because ocean ship trajectories are essentially inside planes, the study's controllers had two degrees of freedom.

3.4.5 Underwater vessels

Since underwater vessels cannot utilize GNSS signals, the most often used sensors in navigation are bathometers, inertial sensors, and Doppler sonar. In underwater transponder positioning (UTP) system, because of its close linkage with the INS, UTP only requires one transponder [43]. A navigation system was created and put into use in a real submarine as well as a simulation to combine information from a bathometer, DVL, and acoustic location system. USBL [Ultra-Short Baseline (used in underwater navigation systems)] is intended for underwater sea-bed modeling applications for oceanic examination, combining DVL, compass, pressure sensor, and IMU.

3.4.6 Outer space

In space, two main categories that include integrated navigation systems exist spacecraft and autonomous satellites. Given its low cost and passive nature, CNS is among the better options for the former. By using the state estimation technique and taking into account the measured locations of celestial bodies, one may estimate the position of a satellite. Its great anti-interference capacity and time-independent location inaccuracy are offset by the horizon sensor's poor precision. Another new technique involves the X-ray pulsar, which contrasts the pulse timing estimates with the actual pulse TOA measured by the X-ray sensors on the satellite. Due to its lone pulsar, its navigation status is unobservable, and its anti-interference capabilities are strong, comparable to those of the central nervous system [46]. Thus, a federated UKF-based pulsar/CNS integrated navigation technique improved location precision, decreased failure risk, and reached velocity accuracy of over 0.1 m/s [47]. Amongst the UKF-based method for combining data for direct and indirect horizon sensing, indirect horizon sensing involves the use of an earth sensor and atmospheric refraction of starlight.

Certain metrics have been employed for spacecraft navigation systems, such as the outputs of the star sensor, the geocentric vector, and the geocentric distance of the earth sensor. Numerous space missions have made on extensive use of the astronavigation system, CNS, and the use of Doppler measurement including KF with enhanced best location and attitude estimates possible by incorporating data from SINS sensor suites and outside assistance, such as guided vehicles' autonomous navigation system. The switch-mode fusing data filter employing Inertial-SLAM (Simultaneous Localization and Mapping) with Unscented Kalman Filter (ISRUKF) and EKF, as well as the iterated square-root UKF filter approach, was given. To roughly represent the true system, this method may integrate a heliocentric vector and range obtained from a navigation sensor with the starlight point of view. It is also feasible to make online modifications to the noise produced by measurement covariance matrices. The sparse-grid phase filter handled relative navigation based on vision for two spacecraft. In addition to gyro biases, this filter provided estimates of relative orbiting and relative attitude.

3.5 FUTURE AVENUES FOR INVESTIGATION

Research has examined a range of integrated navigation systems up to this point, including learning- and analytics-based programs. However, there are still challenges and issues in this sector.

a. **Error Mitigation and Accuracy Assessment:** Accuracy evaluation is usually tough for localization since ground truth is sometimes hard or impossible to achieve, especially in some complex interior contexts like shops, offices, and retail stores. Excessive interference is produced by crowded situations, making it challenging to assess errors with outside support. The accuracy of wireless-based localization is significantly impacted by NLoS and multipath problems. The primary focus of future studies may be on identifying and resolving these problems in integrated systems. By training the historical dataset or by using environment sensing, the identification of multipath and NLoS can be aided by the use of many sensors.
b. **Crowd Source-based Positioning and Collaborative Localization:** As the IoT connects millions of devices and people, enormous volumes of data are being gathered in real time. Thus, Utilizing data from current public infrastructures for collaborative localization and crowd-sourcing is a pattern for inexpensive mass-market IoT applications. The fingerprinting positioning approach primarily saves time and labor in database development because location consumers under the crowd-sourced framework are additionally location suppliers.
c. **Wider Use of Optimization Techniques:** KF-relevant estimating techniques are the mainstay of analytics-based approaches. On the other hand, graph optimization is quite likely to handle nonlinear issues;

moreover, designing an optimization architecture for merging multi-sensor data is not difficult. It is also capable of supporting real-time location and navigation using a sliding window architecture. We expect to see more uses of graph optimization techniques in the future.

d. **Theoretical Features of Sensor Fusion:** Minimizing mistakes is the primary objective of learning-based and analytics-based systems; nonetheless, local minimums rather than worldwide optimization are the common fate of most algorithms. Thus, there is still work to be done on the conversion of a nonconvex problem into a convex one, especially for optimization-based approaches.
e. **INS-Free Filtering Model:** INS error is frequently employed as the state model in KF-relevant systems, making INS an essential component of an integrated navigation system. Future research should examine the state model and INS-free integrated navigation without reliance on INS. In the case of an accelerometer or gyro failure, this will assist in guaranteeing that navigation systems will continue to function. It is necessary to thoroughly analyze the observability of that model.
f. Thorough comparisons among learning and analytics-based strategies show that based on learning navigation systems handle calibration errors and malfunctioning sensors better. Nevertheless, the majority of studies fail to fully distinguish between analytics-based and learning-based approaches, focusing only on accuracy while ignoring other factors like cost and dependability. As a result, choosing the best localization strategy is challenging. A general assessment of these two methods. We thus implore upcoming scholars to focus more on choosing the appropriate methods for the diverse integrated systems. Furthermore, more focus needs to be paid to learning-based techniques like reinforcement learning.
g. **Positioning Using New Sensors:** The development of contemporary technology has led to the creation of several new sensors, many of which may be placed in certain situations. For example, an event-based camera in an urban setting may record moving objects and determine their kinematic properties; a 6G connection combined with high image resolution and sensing enables indoor 10 cm placement.
h. **The Interpretability of Methods based on Learning:** Particularly for deep learning, learning-based techniques are like "black boxes," with difficult-to-understand mechanics. When a training model's interpretability is poor, navigation users are less likely to trust it. This might result in major mishaps when an integrated navigation system lacks resilience. Investigating how deep neural networks learn in the context of multi-sensor fusion is therefore one possible line of inquiry.
i. **Database Variety for Interior Localization:** In the large database age, multisensory navigation research should focus on data variety, particularly for inside localization, rather than on only positioning algorithms. Since people spend eighty percent of their time indoors, deep

learning algorithms may be trained on vast amounts of data. For example, in Location-Based Services (LBS) applications, pedestrian-based localization is crucial. However, basic walking patterns (forward, backward, right, left, and running) were often taken into account by the PDR techniques that were already in use. We anticipate other walking behaviors, such as leaping, running, and sprinting, to obtain more accurate location-based services, future work will consider crouching, stair climbing, and descending steps. We also plan to include more IoT devices—like smart earphones, wristbands, and watches—for pedestrian-based applications to expand the current study's reach.

j. **Integrity Monitoring:** Occasionally, a navigation technology system may generate production errors that are significantly greater than the uncertainty boundaries that are declared or suggested by it. Unusual operating circumstances or malfunctioning hardware or software might be the cause of this. Integrity monitoring systems guard overall navigation solutions by identifying these flaws. The most important navigational attributes are integrity, precision, continuity, and availability. To our knowledge, nevertheless, filtering-based GNSS/INS systems are included in the majority of integrity investigations. Algorithms and other navigation methods have not been thoroughly investigated. In future development, integrity monitoring should receive greater emphasis on constructing a secure multisensory navigation system.

k. **Positioning Privacy:** Previous research often offers an equivalent degree of privacy protection. However, consumers require varying degrees of privacy protection based on factors like the type of data, location, and time of day. For example, the user requests more protection of privacy in the area of interest. On the other hand, the user needs less privacy protection in unimportant areas. A research hotspot is hence a customized location privacy protection. Second, because ignoring existence outputs the preservation of privacy systems trustworthiness, the level of protection of privacy provided by them is underestimated. Furthermore, developing metrics that accurately assess privacy protection technologies remains a difficulty. Thirdly, current studies are rigid and incapable of choosing privacy protection tactics that are situation-specifically appropriate. More significantly, even with the maturation of technologies like blockchain and edge computing, privacy protection remains challenging to implement in a way that strikes a balance between privacy and usefulness. Ultimately, the advent of quantum computing has rendered conventional cryptography-based privacy protection techniques ineffective due to the machine's immense processing capacity. As a result, further research on location privacy protection against quantum assaults is very important.

It is quite likely that integration designs will continue to proliferate over the next ten years, improving in terms of integration, cost, and performance.

3.6 CONCLUSIONS

This chapter thoroughly examines multi-sensor integrated navigation systems, targeting both professionals and academics. It involves a detailed review of existing literature, analyzing various standalone positioning and navigation systems to understand integrated solutions better. These integrated systems are categorized based on three aspects: the sources of data, the algorithms and architectures used, and the scenarios they operate in. To delve deeper, we categorize them further into analytics-based fusion and learning-based fusion, focusing on how they integrate data. Due to the limited discussion on learning-based fusion, we introduce various algorithms like SVM, ANN, deep algorithm learning, random forest reinforcement, and fuzzy logic learning to illustrate their differences and strengths. We stress the importance of considering design before implementation and offer insights into different application scenarios for integrated navigation. Furthermore, we suggest areas for future research to prompt further exploration in this vital field. Ultimately, we anticipate continued advancements in integrated navigation, with broader applications, particularly in domains like the IoT.

3.6.1 Funding

No Research grants from funding agencies.

3.6.2 Declaration of competing interest

The authors declare that they have no known competing financial interests or personal relationships that could have appeared to influence the work reported in this paper.

3.6.3 Availability of data and materials

Not applicable.

REFERENCES

1. T. Zhang, J. Wang, L. Zhang, and L. Guo, "A Student's T-Based Measurement Uncertainty Filter for SINS/USBL Tightly Integration Navigation System," *IEEE Transactions on Vehicular Technology*, vol. 70, no. 9, pp. 8627–8638, 2021, doi:10.1109/TVT.2021.3102085.
2. C. Laoudias, A. Moreira, S. Kim, S. Lee, L. Wirola, and C. Fischione, "A Survey of Enabling Technologies for Network Localization, Tracking, and Navigation," *IEEE Communications Surveys & Tutorials*, vol. 20, no. 4, pp. 3607–3644, 2018, doi:10.1109/COMST.2018.2855063.

3. X. Guo, N. Ansari, F. Hu, Y. Shao, N. R. Elikplim, and L. Li, "A Survey on Fusion-Based Indoor Positioning," *IEEE Communications Surveys & Tutorials*, vol. 22, no. 1, pp. 566–594, 2020, doi:10.1109/COMST.2019.2951036.
4. A. Benini, A. Mancini, and S. Longhi, "An IMU/UWB/Vision-based Extended Kalman Filter for Mini-UAV Localization in Indoor Environment using 802.15.4a Wireless Sensor Network," *Journal of Intelligent & Robotic Systems*, vol. 70, no. 1, pp. 461–476, 2013, doi:10.1007/s10846-012-9742-1.
5. C. Zhang, C. Guo, and D. Zhang, "Data Fusion Based on Adaptive Interacting Multiple Model for GPS/INS Integrated Navigation System," *Applied Sciences*, vol. 8, no. 9, Art. no. 9, 2018, doi:10.3390/app8091682.
6. Q. Wang, Y. Hu, and M. He, "Effect of Filler on the Self-Lubrication Performance of Graphite Antimony Composites," *Journal of China University of Mining and Technology*, vol. 18, no. 3, pp. 441–448, 2008, doi:10.1016/S1006-1266(08)60091-7.
7. D. Loebis, R. Sutton, and J. Chudley, "Review of Multisensor Data Fusion Techniques and their Application to Autonomous Underwater Vehicle Navigation," *Journal of Marine Engineering & Technology*, vol. 1, no. 1, pp. 3–14, 2002, doi:10.1080/20464177.2002.11020159.
8. Y. Zhuang et al., "Multi-Sensor Integrated Navigation/Positioning Systems Using Data Fusion: From Analytics-Based to Learning-Based Approaches," *Information Fusion*, vol. 95, pp. 62–90, 2023, doi:10.1016/j.inffus.2023.01.025.
9. A. Ben-Afia et al., "Review and Classification of Vision-Based Localisation Techniques in Unknown Environments," *IET Radar, Sonar & Navigation*, vol. 8, no. 9, pp. 1059–1072, 2014, doi:10.1049/iet-rsn.2013.0389.
10. K. EltayebYassin and N. I. Elsadig, "Industrial Hygiene and Occupational Safety Assessment in Khartoum North: A Case Study of Food and Beverages Industries," *University of Khartoum Engineering Journal*, vol. 3, no. 1, 2013.
11. R. C. Shit, "Precise Localization for Achieving Next-Generation Autonomous Navigation: State-of-the-Art, Taxonomy and Future Prospects," *Computer Communications*, vol. 160, pp. 351–374, 2020, doi:10.1016/j.comcom.2020.06.007.
12. X. Niu, H. Tang, T. Zhang, J. Fan, and J. Liu, "IC-GVINS: A Robust, Real-Time, INS-Centric GNSS-Visual-Inertial Navigation System," *IEEE Robotics and Automation Letters*, vol. 8, no. 1, pp. 216–223, 2023, doi:10.1109/LRA.2022.3224367.
13. L. Chen, X. Zhou, F. Chen, L.-L. Yang, and R. Chen, "Carrier Phase Ranging for Indoor Positioning With 5G NR Signals," *IEEE Internet of Things Journal*, vol. 9, no. 13, pp. 10908–10919, 2022, doi:10.1109/JIOT.2021.3125373.
14. W. Lechner and S. Baumann, "Global Navigation Satellite Systems," *Computers and Electronics in Agriculture*, vol. 25, no. 1–2, pp. 67–85, 2000.
15. P. Kolar, P. Benavidez, and M. Jamshidi, "Survey of Datafusion Techniques for Laser and Vision based Sensor Integration for Autonomous Navigation," *Sensors*, vol. 20, no. 8, p. 2180, 2020.
16. J. Biswas and M. Veloso, "Wifi localization and navigation for autonomous indoor mobile robots," In: *Presented at the 2010 IEEE International Conference on Robotics and Automation*, IEEE, New York, 2010, pp. 4379–4384.

17. W. Guan et al., "Robot Localization and Navigation Using Visible Light Positioning and SLAM Fusion," *Journal of Lightwave Technology*, vol. 39, no. 22, pp. 7040–7051, 2021.
18. T. Abbasi, R. Sanjeevi, M. Makhija, and S. A. Abbasi, "Role of Vitamins B-3 and C in the Fashioning of Granules in UASB Reactor Sludge," *Applied Biochemistry and Biotechnology*, vol. 167, no. 2, pp. 348–357, May 2012, doi:10.1007/s12010-012-9691-y.
19. P. B. Sathvara, J. Anuradha, R. Sanjeevi, S. Tripathi, and A. B. Rathod, "Spatial analysis of carbon sequestration mapping using remote sensing and satellite image processing," In: *Multimodal Biometric and Machine Learning Technologies*, Wiley Scrivener Publishing LLC, New York, 2023, pp. 71–83. doi:10.1002/9781119785491.ch4.
20. R. Sanjeevi, M. Haruna, S. Tripathi, B. Singh, and A. Jayaraman, "Impacts of Global Carbon Foot Print on the Marine Environment," *International Journal of Engineering Research and Technology*, vol. 5, pp. 51–54, May 2017.
21. M. Zhang, R. Mandraju, U. Rai, T. Shiratsuchi, and M. Tsuji, "Monoclonal Antibodies against Plasmodium falciparum Circumsporozoite Protein," *Antibodies*, vol. 6, no. 3, Art. no. 3, 2017, doi:10.3390/antib6030011.
22. R. C. Shit, S. Sharma, D. Puthal, and A. Y. Zomaya, "Location of Things (LoT): A Review and Taxonomy of Sensors Localization in IoT Infrastructure," *IEEE Communications Surveys & Tutorials*, vol. 20, no. 3, pp. 2028–2061, 2018, doi:10.1109/COMST.2018.2798591.
23. T. Abbasi, R Sanjeevi, J Anuradha, and S. A. Abbasi, "Impact of Al_{3+} on Sludge Granulation in UASB Reactor." https://nopr.niscpr.res.in/handle/123456789/19284 (January, 2013)
24. J. Yan et al., "Transformative Innovations for a Sustainable Future – Part III," *Applied Energy*, vol. 227, pp. 1–6, 2018, doi:10.1016/j.apenergy.2018.07.009.
25. T. Abbasi, J. Anuradha, and S. Abbasi, "Utilization of the Pernicious Aquatic Weed Salvinia (Salvinia molesta DS Mitchell) in Generating Gold Nanoparticles," *Indian Journal of Biotechnology*, vol. 15, pp. 382–391, 2016.
26. T. J. Chong, X. J. Tang, C. H. Leng, M. Yogeswaran, O. E. Ng, and Y. Z. Chong, "Sensor Technologies and Simultaneous Localization and Mapping (SLAM)," *Procedia Computer Science*, vol. 76, pp. 174–179, 2015, doi:10.1016/j.procs.2015.12.336.
27. P. K. Panigrahi and S. K. Bisoy, "Localization Strategies for Autonomous Mobile Robots: A Review," *Journal of King Saud University – Computer and Information Sciences*, vol. 34, no. 8, Part B, pp. 6019–6039, 2022, doi:10.1016/j.jksuci.2021.02.015.
28. D. Wu, Y. Jia, L. Wang, and Y. Sun, "Ship Hull Flexure Measurement based on Integrated GNSS/LINS," *Frontiers of Optoelectronics*, vol. 12, no. 3, pp. 332–340, 2019, doi:10.1007/s12200-019-0867-8.
29. K. Shamaei and Z. M. Kassas, "Receiver Design and Time of Arrival Estimation for Opportunistic Localization With 5G Signals," *IEEE Transactions on Wireless Communications*, vol. 20, no. 7, pp. 4716–4731, 2021, doi:10.1109/TWC.2021.3061985.
30. A. B. Rathod, J. Anuradha, P. B. Sathvara, S. Tripathi, and R. Sanjeevi, "Vegetational Change Detection Using Machine Learning in Gis Technique: A Case Study from Jamnagar (Gujarat)," *Journal of Data Acquisition and Processing*, vol. 38, no. 1, pp. 1046–1061, 2023.

31. J. Zhang et al., “LeGO-LOAM-FN: An Improved Simultaneous Localization and Mapping Method Fusing LeGO-LOAM, Faster_GICP and NDT in Complex Orchard Environments,” *Sensors*, vol. 24, no. 2, p. 551, 2024.
32. P. B. Sathvara, J. Anuradha, Sandeep Tripathi, and R. Sanjeevi, “Impact of climate change and its importance on human performance,” In: A. Chatterjee (ed.), *Insights on Impact of Climate Change and Adaptation of Biodiversity*, 1st ed., KD Publication, Delhi, 2024, pp. 1–9.
33. A. B. Rathod, P. B. Sathvara, A. Tripathi, J. Anuradha, S. Tripathi, and R. Sanjeevi, Vegetational Cartography Analysis Utilizing Multi-temporal NDVI Data Series: A Case Study from Rajkot District (Gujarat), India. *Journal of Tianjing University of Science & Technology*, vol. 55, no. 4, pp. 490–497, 2022.
34. S. Gao, Y. Zhong, X. Zhang, and B. Shirinzadeh, “Multi-Sensor Optimal Data Fusion for INS/GPS/SAR Integrated Navigation System,” *Aerospace Science and Technology*, vol. 13, no. 4, pp. 232–237, 2009, doi:10.1016/j.ast.2009.04.006.
35. V. Amaratunga, L. Wickramasinghe, A. Perera, J. Jayasinghe, and U. Rathnayake, “Artificial Neural Network to Estimate the Paddy Yield Prediction Using Climatic Data,” *Mathematical Problems in Engineering*, vol. 2020, p. 8627824, 2020, doi:10.1155/2020/8627824.
36. A. M. R. Nishimwe and S. Reiter, “Using Artificial Intelligence Models and Degree-Days Method to Estimate the Heat Consumption Evolution of a Building Stock Until 2050: A Case Study in a Temperate Climate of the Northern part of Europe,” *Cleaner and Responsible Consumption*, vol. 5, p. 100069, 2022, doi:10.1016/j.clrc.2022.100069.
37. H. Hong, P. Tsangaratos, I. Ilia, J. Liu, A.-X. Zhu, and W. Chen, “Application of Fuzzy Weight of Evidence and Data Mining Techniques in Construction of Flood Susceptibility Map of Poyang County, China,” *Science of The Total Environment*, vol. 625, pp. 575–588, 2018, doi:10.1016/j.scitotenv.2017.12.256.
38. R. K. Oruganti et al., “Artificial Intelligence and Machine Learning Tools for High-Performance Microalgal Wastewater Treatment and Algal Biorefinery: A Critical Review,” *Science of the Total Environment*, vol. 876, p. 162797, 2023, doi:10.1016/j.scitotenv.2023.162797.
39. M. Lowe, R. Qin, and X. Mao, “A Review on Machine Learning, Artificial Intelligence, and Smart Technology in Water Treatment and Monitoring,” *Water*, vol. 14, no. 9, Art. no. 9, 2022, doi:10.3390/w14091384.
40. A. Malviya and D. Jaspal, “Artificial Intelligence as an Upcoming Technology in Wastewater Treatment: A Comprehensive Review,” *Environmental Technology Reviews*, 2021. https://www.tandfonline.com/doi/abs/10.1080/21622515.2021.1913242
41. Y. Wang et al., “A Review on Applications of Artificial Intelligence in Wastewater Treatment,” *Sustainability*, vol. 15, no. 18, Art. no. 18, 2023, doi:10.3390/su151813557.
42. F. Gu et al., “Indoor Localization Improved by Spatial Context: A Survey,” *ACM Computing Surveys (CSUR)*, vol. 52, no. 3, pp. 1–35, 2019.
43. Ø. Hegrenæs, K. Gade, O. K. Hagen, and P. E. Hagen, “Underwater transponder positioning and navigation of autonomous underwater vehicles,” In: *OCEANS 2009*, Oct. 2009, pp. 1–7. doi:10.23919/OCEANS.2009.5422358.

44. R. C. Simpson, "Smart Wheelchairs: A Literature Review," *Journal of Rehabilitation Research & Development*, vol. 2, no. 6, pp. 1325–1327, 2005.
45. C. Laoudias, A. Moreira, S. Kim, S. Lee, L. Wirola, and C. Fischione, "A Survey of Enabling Technologies for Network Localization, Tracking, and Navigation," *IEEE Communications Surveys & Tutorials*, vol. 20, no. 4, pp. 3607–3644, 2018.
46. H. Kaplan and A. Pyayt, "Fully Digital Audio Haptic Maps for Individuals with Blindness," *Disabilities*, vol. 4, no. 1, pp. 64–77, 2024.
47. M. Bajracharya et al., "Demonstrating Mobile Manipulation in the Wild: A Metrics-Driven Approach," arXiv preprint arXiv:2401.01474, 2024, https://arxiv.org/abs/2401.01474.

Chapter 4

The future of infrastructure production

Exploring low-cost sensor technologies

Rolly Gupta and Lalit Kumar Sagar

4.1 INTRODUCTION

The infrastructure sector is enormous, with a $10 trillion global yearly expenditure. Comparing it to other businesses like manufacturing and agriculture, however, it has been less productive and efficient [1,2]. Using sensing technologies to make the sector more data-oriented is one suggested way to increase output. Sensing technologies hold great promise for completing this necessary data collection and improving quality, productivity, and safety in the infrastructure sector [1]. Sensing technology is revolutionizing data collecting, transmission, and analysis in the infrastructure sector by having the ability to rapidly gather and permanently retain data [1]. Researchers note that while sensors, communications, controls, and analytics technologies are developing quickly, sensor installation and maintenance costs are still relatively high. Furthermore, specialized analysis of the data gathered from these sensors is frequently needed, which can be costly and time-consuming [3–8]. A problem facing the entire business for all applications is the expense of the sensor [4]. The capital expense of implementation has been identified time and time again as a significant obstacle [9]. Other important elements of obstacles include operating, maintenance, and training expenses [10]. The majority of the commercial solutions in use today are highly expensive. These methods use smart devices to collect data, which is subsequently sent across devices and processed collaboratively using cloud computing or other related technologies [9]. The term "low-cost sensors" was not clearly defined by the researchers, nor were the requirements for deeming a sensor unit low-cost established. As a result, the definition provided by the authors themselves is that a low-cost sensor is one that may be used at a lower cost of installation, operation, and maintenance than typical sensors. Although customized analysis may be necessary for the data gathered from inexpensive sensors, they provide an affordable method of data collection. Even while low-cost sensors have been the subject of considerable study, investigations evaluating the accuracy and dependability of various low-cost sensors are still lacking [11].

DOI: 10.1201/9781003633884-4

Low-cost sensors are being used in many industries for a variety of purposes. According to research, the majority of investigations have been conducted on inexpensive sensors that detect soil, marine, geo-hazard, water quantity, water quality, and air quality [12]. Sensing networks and real-time data monitoring help many sectors become more productive and efficient. However, because sensors may measure things like temperature and other generic properties, they need to be adapted before context-specific applications can be investigated [13].

The study is carried out in association with a well-known contracting company that is actively engaged in infrastructure manufacturing. The business recommended a preliminary investigation into inexpensive sensors that might be applied to the manufacturing phase. The goal of the study is to investigate the possible applications of inexpensive sensors in the infrastructure sector's production phase. In the infrastructure phase, the study looks into the state of the art in inexpensive sensors. The primary subjects of this investigation are temperature and humidity sensors. Temperature and humidity were the parameters that the building sector most wanted to be monitored [11]. The majority of the commercial systems that are now on the market are very expensive, and temperature and humidity measurements provide valuable information for numerous applications in civil engineering [14]. Additionally, extant research indicates that there is a study gap in the comparison of inexpensive temperature and humidity sensors [14]. This is due to the current state of the art. The study's scope led to the formulation of the following research questions:

Q1: What is the current state of the art for inexpensive sensors, namely those for temperature and humidity in the infrastructure sector's production phase?

Q2: What possible applications exist for inexpensive temperature and humidity sensors in the infrastructure sector's production stage?

4.2 REVIEW

4.2.1 Literature review

Due to the topic's relative youth and lack of research experience, a qualitative methodology that includes a review and analysis of the literature has been selected for the study. The sections that follow include the specifics of the literature review. There are two aspects to the literature review. The purpose of the first study of the literature is to learn about the history and current situation of low-cost sensors in various industries, and the second review is to learn about the state of the art for low-cost temperature and humidity sensors during the infrastructure manufacturing phase. The sections that follow provide more information on the literature review.

Table 4.1 Details of literature reviews

S. No.	*Topics*	*Description*	*Combination of keywords*
1	Application of low-cost sensors	Low-cost sensors are used in different industries.	"low-cost sensor" AND "application" AND "sector"
2	State of the art of low-cost temperature and humidity sensors in the infrastructure sector	Low-cost temperature and humidity sensors used in the production stage of the infrastructure sector.	"low-cost sensor" AND "Infrastructure sector " AND "temperature" OR "humidity"

4.2.1.1 Current state of inexpensive sensors across several sectors

The search is carried out using the search terms "low-cost sensor," "application," and "sector." Scopus is the database that was used for this search. Only works released between 2019 and 2023 were considered in the literature evaluation in order to concentrate on the most recent and pertinent research. This search aims to obtain a summary of research on low-cost sensors and their applications carried out in various industries. Thus, conference papers, books, and book chapters were not accepted.

Only a small number of articles published in infrastructure were found using the search above. Thus, another search is done using the terms "low-cost sensors," "Infrastructure sector," "temperature," and "humidity" on Scopus and Google Scholar. Table 4.1 displays the combination and description of the keywords that were used. We look through the first hits to find often occurring terms that do not provide value to the study goals. To get rid of results that are not relevant, these terms are utilized as exclusion keywords. The keywords "automobile," "food sector," "agriculture," and "mining" have been recognized as exclusions. We used the articles from 2019 to 2023 for uniformity and relevancy. The literature abstract is screened as the last step in the filtering process. The purpose of this step is to guarantee that only the items from the infrastructure production stage make it into the final list. The initial literature review result also includes the papers that were found through this search.

4.2.2 Interview

The specifics of the conducted interviews are discussed in this section. There were six interviews, and Table 4.2 provides information about the interview subjects. The technique of snowball sampling is used to identify possible candidates. Using a technique called "snowball sampling," the interviewee recommends the following possible choices. The director of the contracting firm's research and development division is interviewed for the first time

Table 4.2 The details of participants in the interview

Participants	*Field*	*Current role*	*Experience in years*
P1	Engineering	Senior researcher	25
P2	Infrastructure	Engineer/firm owner	50
P3	Installation and maintenance	Engineer	10
P4	Microsystems and sensors	Senior expert	25
P5	IT in infrastructure	Researcher/consultant	30
P6	Engineering	Research and development	10

Table 4.3 Theme and description of codes used for the interviews

S. No.	*Theme*	*Description*
1	Definition	What does a low-cost sensor mean?
2	Constraints	What are the main constraints regarding the sensors currently used in the infrastructure sector?
3	Adaptation	What are the key points to be considered for the adaptation of low-cost sensors into infrastructure?
4	Application	What are the potential applications of low-cost temperature and humidity sensors in the infrastructure?

regarding the study. The chosen participants work in fields related to the research topic and have a minimum of five years of experience.

Online interviews were conducted. There was an hourly schedule set for each interview. An interview that was semi-structured was used. A logical method is chosen for the analysis of the interviews. It is a methodical approach to data analysis. There are no predetermined questions because the interview is semi-structured and revolves around four themes: definition, limits, adaptation, and application. Table 4.3 displays a summary of the interview's themes along with their descriptions.

4.3 RESULTS AND ANALYSIS

4.3.1 Literature survey results

4.3.1.1 Application of low-cost sensors in different industries

After a few chosen papers were examined, their data were compiled. Forty-one papers were included in the final review; these were arranged in Table 4.4, according to their applications and study topics; the recurrence of study areas was noted. The particular sector in which the low-cost sensors indicated in the study would be used was taken into

Table 4.4 Applications of low-cost sensors in different industries

S. No.	*Area of study*	*Application*	*References*
1	Food sector	Evaluation of oil's oxidation status, sensitive and selective sensing approach, supply chain monitoring, speeding up quality control, intelligent food packaging, speeding up quality control	[15–19]
2	Infrastructure sector	Corrosion monitoring, particulate matter monitoring, structural health monitoring, measuring dust level at site, monitoring crack	[20–24]
3	Mining sector	Space–time monitoring of PM concentrations in mines, 3D surface profiling	[25,26]
4	Agricultural sector	Sensitive and selective sensing approach, supply chain monitoring, distance or length measurement, high-throughput phenotyping, crop surface model	[16,17,27–29]
5	Pharmaceutical/ health industries	Sensitive and selective sensing approach, biomedical application, monitoring safety of fragile items, airbag deployment, health monitoring, weight detection	[16,30–32]
6	Mechanical sector	Tool condition monitoring, wear prediction, preprocessing monitoring data from complex systems aimed for DDMs, damage detection, distance ranging, proximity sensing and flow measurement	[33–36]
7	Manufacturing sector	Textile recognition, predictive maintenance, using IR sensors in HVAC, precision manufacturing, preprocessing monitoring data from complex systems aimed for DDMs, ethanol sensing and detection, measuring distance	[33,37–39,34,40]
8	Environmental area	Air quality monitoring, measure water quantity, indoor physical environment	[41–47]
9	Chemical sector	Gas detection, low power-consuming cost-efficient VOC sensors	[48,49]
10	Automobile sector	Predictive maintenance, air quality monitoring, using IR sensors in HVAC, monitoring safety of fragile items, airbag deployment, health monitoring, weight detection, navigation applications, measuring distance, distance ranging, proximity sensing and flow measurement, positioning and navigation, 3D surface profiling	[38,42,17,50,51,32, 52,40,36,53,26]
11	Communication sector	IAQ monitoring, digital calibration certificates	[54,55]

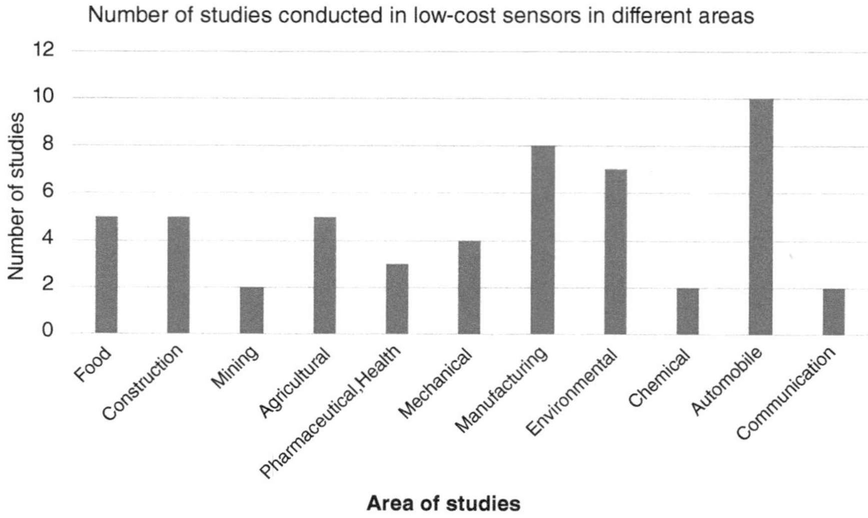

Figure 4.1 Number of studies conducted on low-cost sensors in different areas.

consideration when determining the study area or sector. Figure 4.1 displays the analytical result in graphical form. The review's findings showed that the majority of research has concentrated on their use in the automotive ($n=10$) and environmental ($n=7$) domains. The manufacturing sector has eight research projects. Five papers that address low-cost sensor applications in the sector's infrastructure phase were uncovered in this literature review.

4.3.1.2 State-of-the-art low-cost temperature and humidity sensors in infrastructure sector

Five papers that addressed the use of inexpensive sensors in the infrastructure sector were found through the literature review. As indicated in Table 4.5, these studies were examined to determine the kind of sensors that were discussed in the study. Applications include dust level measurement at the site, corrosion monitoring, structural health monitoring, and particulate matter monitoring. Cracks in the structures are also monitored. Only two of these five studies address the application of temperature and humidity sensors during the infrastructure production process. Because fluctuations in temperature and humidity levels can exacerbate corrosion, one study focuses on corrosion monitoring and uses temperature and humidity sensors. Low-cost sensors are used to assess temperature, humidity, pressure, and crack breadth in the second study, which deals with structural health monitoring.

Table 4.5 Applications and types of sensors used in the production stage

Application	*Type of sensor*	*References*
Corrosion monitoring	Electrochemical sensors, temperature, humidity sensors	[33]
Air quality monitoring, particulate matter monitoring	Particulate matter sensor	[21]
Structural health monitoring	Temperature, humidity, pressure, crack width, etc.	[22]
To measure dust level at site	Dust sensor	[23]
To monitor crack	Piezoelectric sensor	[24]

4.3.1.3 Potential applications of low-cost temperature and humidity sensors

Potential uses for temperature and humidity sensors are discussed in this section based on published research. Through the process of back-citing, which entails examining the references cited in the literature that was retrieved, these applications were found. To ascertain the precise uses, kinds of sensors used, and techniques used, the research were examined. The purpose of this analysis was to determine if the investigations were carried out as case studies or in laboratory environments. By looking at these variables, it determines the extent to which the sector has established sensor utilization.

- **Interview:** The findings of the interview that was done as part of the study are shown in this section. The interview answers are divided into four portions, each of which has the answers related to a different theme.
- **Definition:** This code was created in order to obtain a potential definition of a low-cost sensor, as the topic was not sufficiently covered in the body of current literature. According to P1, a sensor is deemed low-cost if it costs no more than 1,000 INR each. P2 on the other hand suggested that it be simple to mount and operate with a reasonable level of accuracy and precision. Additionally, inexpensive sensors that are reusable are recommended for concrete curing and monitoring. P3 recommended that cordless and easily movable sensors be used. P4 contended that the cost of a "low-cost" sensor can range from 50% to 100%, depending on the product and application or more affordably priced. P5 claimed that the low-cost sensor's component parts ought to be standardized. On the other hand, P6 recommended that the most important factor be stopped and correct data be monitored continuously across time. The monitoring system would be considered low-cost if the cost per unit of measurement was less than 1,000 INR.

- **Constraints:** P1 proposed that the two main limitations on sensors are the necessity for a large number of data points and the high cost of storing the data that are gathered. P2 asserted that wireless sensors on the market are not inexpensive and that wired sensors are unsuitable for infrastructure sites. According to P3, the outside sensor may occasionally need a weather shield, which could raise the price and be viewed as a financial limitation. P5 brought up the issue of a lack of readily available, reasonably priced, and manageable sensors for infrastructure sites. P6 presented a similar argument, stating that robust and long-lasting low-cost wireless sensor devices are hard to come by.
- **Adaptation Factors:** P1 suggested that the thickness or size of the sensor unit matters for the adaption factors inquiry. The possibility of reusing the sensor should also be taken into account. P2 had a similar view, stating that if the sensor is placed inside the concrete, its ability to connect with other components is a crucial consideration. P3 recommended that the sensor's interoperability with other gear be taken into consideration. The sensor must also meet the customer's standards and specifications, and maintenance should be taken into account. P4 asserted that several aspects of the sensor system's packing and assembly, such as how wireless it is, how it communicates, how it is powered, and how much data it can deliver, must be considered. P5 suggested that the sensor be developed to function at larger-scale sites as well as lab settings. Analyzing data gathered from the location and contrasting it with laboratory data ought to be feasible. The design should be such that it will endure hard environments and rough handling, even if P6 advised that the sensors and equipment be sturdy and infrastructure-site-friendly.
- **Application:** The following are the primary applications that were gathered from the interviews:
    - Keeping an eye on the quality of the concrete mix: The efficiency of the concrete and the quality of the concrete mix can be studied by keeping an eye on the mixing process and gathering data on temperature and humidity levels. This operation is still typically carried out using destructive means, which is the traditional manner. The participants think that the creation of inexpensive sensors will provide more effective and digital data monitoring and analysis.
    - Automation of the curing process for concrete: The data can be tracked and notified directly to the supervisor's mobile phone or directly to the monitoring devices during the concrete placing and curing process. This data allows for the regulation of the entire curing process. For instance, it allows for the precise calculation of the water required for curing, which helps to maintain the ideal

environmental conditions. The panelists propose that the infrastructure process will become more automated at a faster pace with the use of inexpensive sensors.
- Determining the ideal moment to remove formworks: The information gathered from temperature and humidity sensors will be useful in determining when concrete formworks should be removed.

4.4 DISCUSSION

The state of the art for inexpensive temperature and humidity sensors during the Infrastructure's production phase was determined with the aid of the literature review. The use of inexpensive temperature and humidity sensors during the Infrastructure's production phase was covered in just two publications. The analysis's conclusions support the research gap that has been identified in the publications, which states that despite the wide range of potential uses for inexpensive sensors, these applications have not been explored. This is significant because studies have indicated that the best factors to monitor in the building sector are humidity and temperature [11,14]. Additionally, compared to the infrastructure sector, the data indicate that there have been more research conducted in other industries like manufacturing and automobiles.

During the interviews, the participants expressed varying viewpoints regarding what constitutes low-cost sensors. Combining the viewpoints of those interviewed, a thorough description of inexpensive sensors utilized in the manufacturing stage of the following would be the infrastructure: These sensors maintain a moderate level of accuracy and precision, are wireless and portable, facilitate component standardization, are less expensive than traditional sensors, and continuously gather data. Based on the information gathered from the interviews, the primary obstacle to the deployment of sensors during the infrastructure phase is the lack of accessible, affordable wireless sensors. According to the interviewees, sensors meant for use in infrastructure should be more resilient and able to endure the harsh environmental conditions that are frequently seen at infrastructure sites. Research suggests that wired sensors may experience cable failure as a result of exposure to the environment or possible damage from extreme events, which supports the requirement for wireless sensors at infrastructure sites. The interviews have also shown that there may be financial restrictions related to the expense of storing the sensor data as well as the need to cover or protect outdoor sensors outside.

The literature retrieved from the back citation and the interview results are comparable in terms of possible applicability. The low-cost temperature and humidity sensors indicated in Table 4.6 have potential uses in processes

Table 4.6 The potential applications of low-cost temperature and humidity sensors reported in literature

S. No.	*Application*	*Type of sensor*	*Type of study*	*References*
1	To record the concrete hardening process in an early phase, which can be used to evaluate the quality, compressive strength, and flatness of concrete	Temperature, humidity	Laboratory study	[1,56]
2	Strength of concrete structure with time	Temperature	Laboratory study	
3	Plastic shrinkage cracks, drying shrinkage cracks/cracks in structural components	Temperature, humidity	Laboratory study and field case study	
4	Shrinkage crack monitoring for mass concrete infrastructure/ crack detection in dams	Temperature	Field case study	[7]
5	To decide the minimum striking time of vertical formwork	Temperature, humidity	Laboratory study	
6	Concrete curing efficiency	Temperature, humidity	Laboratory study and field case study	
7	Concrete mix quality	Temperature, humidity	Laboratory study	
8	Real-time monitoring of the on-site persons/real-time monitoring of physical fatigue in infrastructure workers/to check the overexposure to humidity and temperature of workers which could cause occupational diseases and increase the risk of infrastructure safety accidents	Temperature, humidity	Theoretical paper with a field case study	
9	Assisted management with winter infrastructure and freezing method Infrastructure	Temperature	Laboratory study	
10	Structural health monitoring for infrastructure safety management	Temperature	Theoretical paper with laboratory case study	

that enhance or guarantee the safety of infrastructure, materials, and processes. The process and material quality will be improved by applications such as measuring concrete strength, crack monitoring, figuring out curing efficiency, and assessing the quality of the concrete mix. Infrastructure safety management includes applications like structural health monitoring, real-time worker monitoring, and infrastructure freezing. Adoption of sensing technologies will improve infrastructure processes in terms of safety, productivity, and quality performance, according to existing literature [1]. According to recent studies, the use of sensors in automated curing systems can reduce costs by 71.5% and water consumption by 51.2%. By regulating the temperature and relative humidity of the concrete, it is even possible to completely eradicate early-age cracks. Wearable sensor technology is available to monitor temperature, humidity, and other risk factors for infrastructure workers' safety, hence promoting occupational safety [1]. It is interesting to notice that the majority of the prospective applications stated in the literature are carried out in laboratories. There is a dearth of information regarding sensing technologies in actual projects [1].

The interviewees brought out the issue of the concrete laying and curing stages' lack of digitization. They proposed that the use of inexpensive sensors would increase the efficiency and quality of these procedures. Furthermore, digitalization is still very rare in the production phase, even with improvements in the infrastructure sector and a rise in digitalization in the building operation and maintenance phases. This is further supported by applications found in the literature. For instance, traditional methods are still employed for concrete processes like curing, despite the fact that concrete is regarded as the primary and most important infrastructure material in the infrastructure sector.

The research has limitations because the results and literature review analysis were conducted using papers that were sourced from just two databases: Scopus and Google Scholar. Furthermore, the literature evaluation focused exclusively on two kinds of inexpensive sensors that are employed throughout the infrastructure production phase. A constrained collection of questions that matched the study questions was chosen for the interview. There were just six interviewers, all of whom were either employed or descended from India.

4.5 CONCLUSIONS

The literature evaluation results have shown that there is a research deficit in the field of low-cost sensors used in infrastructure production. This offers a chance for more research in this area, which can improve the automation and digitalization of infrastructure procedures. The study's applications have the ability to raise infrastructure safety standards and increase the caliber of the materials and processes employed. Confirming the low-cost sensors' suitability for the study's identified applications will require additional investigation. The findings of the interview indicate that inexpensive sensor solutions are required to hasten the automation of infrastructure procedures.

REFERENCES

1. Arabshahi M, Wang D, Sun J, Rahnamayiezekavat P, Tang W, Wang Y, Wang X. Review on Sensing Technology Adoption in the Infrastructure Sector. *Sensors*. 2021; 21(24):8307. https://doi.org/10.3390/s21248307
2. Barbosa F, Woetzel J, Mischke J. *Reinventing Infrastructure: A Route of Higher Productivity*. 2017, McKinsey Global Institute.
3. Ferrari S, Beccali M. Energy-environmental and cost assessment of a set of strategies for retrofitting a public building toward nearly zero-energy building target. *Sustainable Cities and Society*. 2017; 32:226–34. https://doi.org/10.1016/j.scs.2017.03.010
4. Bursill J. An approach to data-driven sensing and predictive supervisory control for commercial buildings with in-situ evaluation (Doctoral dissertation, Carleton University, 2020). https://doi.org/10.22215/etd/2020-14103
5. Wang B, Xia X. Optimal maintenance planning for building energy efficiency retrofitting from optimization and control system perspectives. *Energy and Buildings*. 2015; 96:299–308. https://doi.org/10.1016/j.enbuild.2015.03.032
6. Killian M, Kozek M. Ten questions concerning model predictive control for energy efficient buildings. *Building and Environment*. 2016; 105:403–12. https://doi.org/10.1016/j.buildenv.2016.05.034
7. Ruparathna R, Hewage K, Sadiq R. Improving the energy efficiency of the existing building stock: A critical review of commercial and institutional buildings. *Renewable and Sustainable Energy Reviews*. 2016; 53:1032–45. https://doi.org/10.1016/j.rser.2015.09.084
8. Minoli D, Sohraby K, Occhiogrosso B. IoT considerations, requirements, and architectures for smart buildings: Energy optimization and next-generation building management systems. *IEEE Internet of Things Journal*. 2017; 4(1):269–83. https://doi.org/10.1109/JIOT.2017.2647881
9. Alizadehsalehi S, Yitmen I. A concept for automated Infrastructure progress monitoring: Technologies adoption for benchmarking project performance control. *Arabian Journal for Science and Engineering*. 2019; 44:4993–5008. https://doi.org/10.1007/s13369-018-3669-1
10. Goodrum PM, Haas CT, Caldas C, Zhai D, Yeiser J, Homm D. Model to predict the impact of a technology on infrastructure productivity. *Journal of Infrastructure Engineering and Management*. 2011; 137(9):678–88. https://doi.org/10.1061/(ASCE)CO.1943-7862.0000328
11. Pereira PF, Ramos NM. Low-cost Arduino-based temperature, relative humidity, and CO_2 sensors: An assessment of their suitability for indoor built environments. *Journal of Building Engineering*. 2022; 60:105151. https://doi.org/10.1016/j.jobe.2022.105151
12. Mao F, Khamis K, Krause S, Clark J, Hannah DM. Low-cost environmental sensor networks: Recent advances and future directions. *Frontiers in Earth Science*. 2019; 7:221. https://doi.org/10.3389/feart.2019.00221
13. Linderoth HCJ. Fiery spirits and supporting programs of action: Keys to exploration and exploitation of open technologies. *International Journal of Healthcare Technology and Management*. 2002; 4:319–332. https://doi.org/10.1504/IJHTM.2002.001146

14. Mobaraki B, Komarizadehasl S, Castilla Pascual FJ, Lozano Galant JA. Environmental monitoring system based on low-cost sensors. In: *Current Topics and Trends on Durability of Building Materials and Components: Proceedings of the XV Edition of the International Conference on Durability of Building Materials and Components (DBMC 2020)*, Barcelona, Spain, 20–23 October 2020 (pp. 1679–1686). https://dx.doi.org/10.23967/dbmc.2020.201
15. Osheter T, Campisi Pinto S, Randieri C, Perrotta A, Linder C, Weisman Z. Semi-autonomic Ai LF-NMR sensor for industrial prediction of edible oil oxidation status. *Sensors*. 2023; 23(4):2125. https://doi.org/10.3390/s23042125
16. Lopes IS, Cassas F, Veiga TA, de Oliveira Silva FR, Courrol LC. Synthesis and characterization of Eugenia uniflora L. silver nanoparticles and l-cysteine sensor application. *Chemistry & Biodiversity*. 2023; 20(1):e202200787. https://doi.org/10.1002/cbdv.202200787
17. Pincheira M, Vecchio M, Giaffreda R. Characterization and costs of integrating blockchain and IoT for agri-food traceability systems. *Systems*. 2022; 10(3):57. https://doi.org/10.3390/systems10030057
18. Cecchini C, Antonucci F, Costa C, Marti A, Menesatti P. Application of near-infrared handheld spectrometers to predict semolina quality. *Journal of the Science of Food and Agriculture*. 2021; 101(1):151–7. https://doi.org/10.1002/jsfa.10625
19. Linderoth HCJ Fiery spirits and supporting programs of action: Keys to exploration and exploitation of open technologies. *International Journal of Healthcare Technology and Management*. 2002; 4:319–332. https://doi.org/10.1504/IJHTM.2002.001146
20. Komary M, Komarizadehasl S, Tošić N, Segura I, Lozano-Galant JA, Turmo J. Low-cost technologies used in corrosion monitoring. *Sensors*. 2023; 23(3):1309. https://doi.org/10.3390/s23031309
21. Koltes JE, Cole JB, Clemmens R, Dilger RN, Kramer LM, Lunney JK, McCue ME, McKay SD, Mateescu RG, Murdoch BM, Reuter R. A vision for development and utilization of high-throughput phenotyping and big data analytics in livestock. *Frontiers in Genetics*. 2019; 10:1197. https://doi.org/10.3389/fgene.2019.01197
22. Mishra M, Lourenço PB, Ramana GV. Structural health monitoring of civil engineering structures by using the internet of things: A review. *Journal of Building Engineering*. 2022; 48:103954. https://doi.org/10.1016/j.jobe.2021.103954
23. Paolucci R, Rotilio M, Ricci S, Pelliccione A, Ferri G. A sensor-based system for dust containment in the infrastructure site. *Energies*. 2022; 15(19):7272. https://doi.org/10.3390/en15197272
24. Du C, Yang Y, Hordijk DA. Experimental investigation on crack detection using imbedded smart aggregate. In: Du C, Yang Y, Hordijk DA (eds.), *Life Cycle Analysis and Assessment in Civil Engineering: Towards an Integrated Vision*, 2018 October 31 (pp. 1199–1206). CRC Press.
25. Zafra-Pérez A, Boente C, de la Campa AS, Gómez-Galán JA, de la Rosa JD. A novel application of mobile low-cost sensors for atmospheric particulate matter monitoring in open-pit mines. *Environmental Technology & Innovation*. 2023; 29:102974. https://doi.org/10.1016/j.eti.2022.102974

26. Popa A, Hnatiuc M, Paun M, Geman O, Hemanth DJ, Dorcea D, Son LH, Ghita S. An intelligent IoT-based food quality monitoring approach using low-cost sensors. *Symmetry*. 2019; 11(3):374. https://doi.org/10.3390/sym11030374
27. Beyaz A, Gerdan D. Potato classification by using ultrasonic sensor with LabVIEW. *Agricultural Science Digest-A Research Journal*. 2020; 40(4): 376–81. https://doi.org/10.18805/ag.D-173
28. Belton D, Helmholz P, Long J, Zerihun A. Crop height monitoring using a consumer-grade camera and UAV technology. *PFG–Journal of Photogrammetry, Remote Sensing and Geoinformation Science*. 2019; 87(5):249–62. https://doi.org/10.1007/s41064-019-00087-8
29. Wilson S, Potgieter J, Arif KM. Robot-assisted floor surface profiling using low-cost sensors. *Remote Sensing*. 2019; 11(22):2626. https://doi.org/10.3390/rs11222626
30. Vyas AK, Gangwar RK, Kumar S. Elliptical air hole PCF-based low-cost sensor for refractive index and temperature detection: Design and analysis. *Optical Fiber Technology*. 2022; 73:103060. https://doi.org/10.1016/j.yofte.2022.103060
31. Narasimman S, Balakrishnan L, Alex ZC. Clad-modified fiber optic sensor utilizing CdS nanoflower as cladding for the detection of ethanol. *Journal of Materials Science: Materials in Electronics*. 2021; 32(19):23900-10. https://doi.org/10.1007/s10854-021-06843-0
32. Tripathy AR, Choudhury A, Dash A, Panigrahi P, Kumar SS, Pancham PP, Sahu SK, Mallik S. Polymer matrix composite engineering for PDMS based capacitive sensors to achieve high-performance and broad- range pressure sensing. *Applied Surface Science Advances*. 2021; 3:100062. https://doi.org/10.1016/j.apsadv.2021.100062
33. Qiang B, Shi K, Liu N et al. Application of cutting power consumption in tool condition monitoring and wear prediction based on Gaussian process regression under variable cutting parameters. *The International Journal of Advanced Manufacturing Technology*. 2023; 124:37–50. https://doi.org/10.1007/s00170-022-10459-x
34. Cofre-Martel S, Lopez Droguett E, Modarres M. Big machinery data preprocessing methodology for data- driven models in prognostics and health management. *Sensors*. 2021; 21(20):6841. https://doi.org/10.3390/s21206841
35. Aguero M, Maharjan D, Rodriguez Md P, Mascarenas DDL, Moreu F. Design and implementation of a connection between augmented reality and sensors. *Robotics*. 2020; 9(1):3. https://doi.org/10.3390/robotics9010003
36. Kang L, Feeney A, Dixon S. The high frequency flexural ultrasonic transducer for transmitting and receiving ultrasound in air. *IEEE Sensors Journal*. 2020; 20(14):7653–60. https://doi.org/10.1109/JSEN.2020.2981667
37. Lejarreta-Andrés J, Melià-Seguí J, Bhattacharyya R, Vilajosana X, Sarma SE. Toward low-cost RF-based bulk fabric classification for the textile sector. *IEEE Sensors Journal*. 2022; 22(16):16586–94. doi: 10.1109/JSEN.2022.3188936.
38. Dorst T, Gruber M, Seeger B, Vedurmudi AP, Schneider T, Eichstädt S, Schütze A. Uncertainty-aware data pipeline of calibrated MEMS sensors used for machine learning. *Measurement: Sensors*. 2022; 22:100376. https://doi.org/10.1016/j.measen.2022.100376

39. Iqbal A, Mian NS, Longstaff A, Fletcher S. Performance evaluation of low-cost vibration sensors in industrial IoT applications. *International Journal of Automation Technology*. 2022; 16(3):329–39. https://doi.org/10.20965/ijat.2022.p0329
40. Bertacchini A, Lasagni M, Sereni G. Effects of the target on the performance of an ultra-low power eddy current displacement sensor for industrial applications. *Electronics*. 2020; 9(8):1287. https://doi.org/10.3390/electronics9081287
41. Siouti E, Skyllakou K, Kioutsioukis I, Patoulias D, Fouskas G, Pandis SN. Development and application of the SmartAQ high-resolution air quality and source apportionment forecasting system for European urban areas. *Atmosphere*. 2022; 13(10):1693. https://doi.org/10.3390/atmos13101693
42. Arroyo P, Gómez-Suárez J, Herrero JL, Lozano J. Electrochemical gas sensing module combined with unmanned aerial vehicles for air quality monitoring. *Sensors and Actuators B: Chemical*. 2022; 364:131815. https://doi.org/10.1016/j.snb.2022.131815
43. Pamula ASP, Ravilla A, Madiraju SVH. Applications of the internet of things (IoT) in real-time monitoring of contaminants in the air, water, and soil. *Engineering Proceedings*. 2022; 27(1):26. https://doi.org/10.3390/ecsa-9-13335
44. Gobatti L, Martins JR, Pereira MC, Leite BC. Real-time sensing and low-cost experimental setup for water quantity investigation in Nature-based Solutions. *Blue-Green Systems*. 2022; 4(2):75–88. https://doi.org/10.2166/bgs.2022.007
45. Shi B, Catsamas S, Kolotelo P, Wang M, Lintern A, Jovanovic D, Bach PM, Deletic A, McCarthy DT. A low-cost water depth and electrical conductivity sensor for detecting inputs into urban stormwater networks. *Sensors*. 2021; 21(9):3056. https://doi.org/10.3390/s21093056
46. Saini J, Dutta M, Marques G. Sensors for indoor air quality monitoring and assessment through Internet of Things: A systematic review. *Environmental Monitoring and Assessment*. 2021; 193(2):66. https://doi-org.proxy.library.ju.se/10.1007/s10661-020-08781-6
47. Nuñez J, Koldeweij R, Trimboli J, Boersma A. Chemical identifier for particulate matter monitoring in infrastructure sites. In: *2021 IEEE Sensors, 2021* October 31 (pp. 1–4). IEEE. https://doi.org/10.1109/SENSORS47087.2021.9639672
48. Narkhede P, Walambe R, Chandel P, Mandaokar S, Kotecha K. MultimodalGasData: Multimodal dataset for gas detection and classification. *Data*. 2022; 7(8):112. https://doi.org/10.3390/data7080112
49. Kiaee MM, Maeder T, Brugger J. Inkjet-printed composites for room-temperature VOC sensing: From ink formulation to sensor characterization. *Advanced Materials Technologies*. 2021; 6(1):2000929. 42. https://doi-org.proxy.library.ju.se/10.1002/admt.202000929
50. Altaf MA, Ahn J, Khan D, Kim MY. Usage of IR sensors in the HVAC systems, vehicle and manufacturing industries: A review. *IEEE Sensors Journal*. 2022. https://doi.org/10.1109/JSEN.2022.3166190
51. Cha Y, Yin C, Du J, Xia T, An W, Zhang S. Real-driving measurement of vehicle interior air quality and cabin air filtering performance by using low-cost sensors. *SAE International Journal of Commercial Vehicles*. 2022; 15. https://doi.org/10.4271/02-15-04-0020

52. El-Sheimy N, Youssef A. Inertial sensors technologies for navigation applications: State of the art and future trends. *Satellite Navigation*. 2020; 1(1):1–21. https://doi.org/10.1186/s43020-019-0001-5
53. Kitani T. Bikeinformatics: An introduction of informatics to the motorcycle researches and the development of new generation motorcycle-based personal vehicles. *Journal of Information Processing*. 2020; 28:3–15. https://doi.org/10.2197/ipsjjip.28.3
54. Xian C, Zhang D, Dai C, Wang CC. Fast generation of high-fidelity RGB-D images by deep learning with adaptive convolution. *IEEE Transactions on Automation Science and Engineering*. 2020; 18(3):1328–40. 37. https://doi.org/10.1109/TASE.2020.3002069
55. Mustapää T, Nikander P, Hutzschenreuter D, Viitala R. Metrological challenges in collaborative sensing: Applicability of digital calibration certificates. *Sensors*. 2020; 20(17):4730. https://doi.org/10.3390/s20174730
56. John ST, Roy BK, Sarkar P, Davis R. IoT enabled real-time monitoring system for early-age compressive strength of concrete. *Journal of Infrastructure Engineering and Management*. 2020; 146(2):05019020. https://doi.org/10.1061/(ASCE)CO.1943-7862.0001754

Chapter 5

Smart sensors for smart manufacturing

Shimpy Sarkar

5.1 INTRODUCTION TO SMART MANUFACTURING

In the dynamic and ever-changing realm of contemporary industry, smart manufacturing emerges as a groundbreaking methodology that incorporates state-of-the-art technologies to enhance production procedures and foster unparalleled levels of effectiveness, adaptability, and ingenuity. Smart production utilizes modern digital technology, automation, data analytics, and networked systems to establish a highly intelligent and interconnected ecosystem, in contrast to conventional production methods that heavily depend on manual labor and static processes. Intelligent systems and gadgets, commonly referred to as "smart" technology, form the core of smart manufacturing (Mittal et al., 2019). The aforementioned components encompass a diverse array of elements, including sensors, actuators, robots, artificial intelligence (AI), Internet of Things (IoT) devices, cloud computing, and data analytics tools. These technologies operate in conjunction to gather, analyze, and respond to extensive quantities of real-time data, empowering firms to make decisions based on data, enhance production processes, and promptly adapt to evolving market requirements. The widespread adoption of smart sensors is a crucial factor in facilitating smart manufacturing. In the production environment, sensors are strategically placed to collect essential data pertaining to equipment performance, product quality, ambient conditions, and supply chain logistics (Sheth and Kusiak, 2022). Manufacturers are able to achieve unparalleled visibility into their operations, resulting in higher productivity, decreased downtime, improved product quality, and ultimately, increased revenue, through the utilization of smart sensors. In the chapter, we will extensively explore the domain of smart manufacturing, with a specific emphasis on the function and influence of smart sensors. This study aims to examine the diverse categories of smart sensors, their utilization in various industrial domains, the advantages they offer to manufacturing procedures, and the obstacles and prospects associated with their implementation (Yao et al., 2019). Our objective is to offer a thorough examination of how smart sensors are revolutionizing the manufacturing industry and facilitating a more intelligent

DOI: 10.1201/9781003633884-5

and adaptable future. This will be achieved by utilizing real-life illustrations, case studies, and analysis of developing trends.

5.1.1 The origins of smart manufacturing

The origins of present-day manufacturing can be traced back to the preceding 50 years. Automation in manufacturing has been facilitated by advancements in computer and machine-building technologies. In contemporary times, machine tools predominantly rely on computer programs as opposed to human operators. Automated material handling systems are utilized for the transportation of materials and components, while automated storage and retrieval systems are employed for their storage.

The degree to which a manufacturing floor is mechanized and the degree to which diverse functional production areas are integrated determine the name given to automated manufacturing since the 1980s. A few examples of these terms are intelligent manufacturing, computer-integrated manufacturing, flexible manufacturing cells, and flexible manufacturing systems. Around 1990, when the *Journal of Intelligent Manufacturing* was founded (Kusiak, 1990) and *Intelligent Manufacturing Systems* was published (Kusiak, 1990), a book that had been in the works for a while prior to its publication, the most current terminology was introduced. Simultaneously, Japan initiated research in the field of intelligent manufacturing, resulting in the development of the Intelligent Manufacturing System (IMS) Program in 1995. This program was designed to provide funding for industry research in Japan. It was recognized that the manufacturing sector of a single country alone was insufficient to transform the industry, and that international collaboration was necessary. Notably, Japan stands out as the leading participant, boasting the highest number of actively engaged corporations. In subsequent years, the IMS Program underwent expansion, as the European Union initiated research endeavors in the field of intelligent manufacturing (Groumpos, 1995). The field of manufacturing is undergoing continuous evolution, resulting in its manifestation in various forms. The manufacturing community has shown increasing interest in the concept of the IoT in recent years. Smart manufacturing incorporates various architectural and technological principles, including cyber-physical systems, the IoT, cloud computing, service-oriented computing, AI, and data science. Once put into practice, these intersecting ideas and technologies will establish manufacturing as the defining characteristic of the upcoming industrial revolution.

5.1.2 Overview of smart manufacturing

Smart manufacturing signifies a fundamental transformation in the operational practices of industries, as it incorporates cutting-edge technologies and data-driven analysis to optimize efficiency, productivity, and adaptability throughout every step of the manufacturing process. The utilization of networked systems, real-time data analytics, automation, and intelligent decision-making in this disruptive approach enables the optimization of operations, waste reduction, quality improvement, and acceleration of innovation. Smart manufacturing is defined by the integration of digital technologies with conventional manufacturing methods, resulting in a dynamic environment where machines, devices, and humans work together smoothly. The aforementioned collaboration facilitates a proactive and adaptable strategy toward production, wherein the use of predictive maintenance, real-time monitoring, and continual optimization leads to enhancements in performance and reductions in costs.

5.1.2.1 *The fundamental elements of smart manufacturing encompass*

The Premise of the IoT and its Connection: The production environment is equipped with IoT devices, including sensors, actuators, and smart devices, which gather and communicate data in real time. The interconnectivity of this network allows for the smooth transmission of information among equipment, systems, and stakeholders, hence facilitating decision-making based on data and optimizing processes (Qu et al., 2019).

Analysis and Utilization of Large-Scale Data and AI: The processing and analysis of the vast amounts of data produced by IoT devices involve the utilization of sophisticated analytics and AI techniques. The utilization of a data-driven methodology enables manufacturers to get valuable insights, detect patterns, and establish correlations (Zheng et al., 2018). This empowers them to determine opportunities for optimization, forecast maintenance requirements, manage inventory levels, and improve overall operational performance.

Automation and Robotics: Automated systems and robotics are key components of smart manufacturing, since they are responsible for executing repetitive activities, assembly processes, and material handling operations with a high degree of accuracy and effectiveness. Collaborative robots, sometimes referred to as cobots, operate in conjunction with human operators, thereby augmenting productivity and safety, while concurrently diminishing cycle times and costs.

Digital Twins and Simulation: The utilization of digital twin technology enables the creation of virtual replicas of tangible assets, processes, and systems. This facilitates manufacturers in simulating and enhancing production situations within a devoid of risk. Through the examination of digital twin data, manufacturers have the ability to refine processes, experiment with novel tactics, and enhance resource allocation in order to achieve enhanced outcomes (Fathi-moghaddam et al., 2018).

Integration and Optimization of the Supply Chain: Smart manufacturing is based on the principle of optimizing and integrating the whole supply chain, from the procurement of raw materials to the shipment of finished goods (Edgar and Pistikopoulos, 2018). The establishment of integration among suppliers, logistics providers, and customers facilitates the smooth coordination, immediate visibility, and adaptable reaction to variations in demand, disruptions, and shifts in the market.

Ongoing Enhancement and Originality: The proliferation of connection and digitization necessitates the implementation of strong cybersecurity protocols in order to safeguard sensitive data, intellectual property, and operational infrastructure. In order to ensure the protection of assets and foster confidence among stakeholders, smart manufacturing programs place a high emphasis on the implementation of cybersecurity procedures, encryption standards, access restrictions, and data privacy regulations (Periera et al., 2017).

Continuous Improvement and Innovation: Smart manufacturing is characterized by its inherent iterative and adaptive nature, which is fostered by a culture that emphasizes continual development and innovation. In order to discover areas for improvement, apply best practices, and remain at the forefront of industry trends and emerging technologies, manufacturers adopt feedback loops, performance measurements, and benchmarking.

5.1.2.2 Pillars of smart manufacturing

The ideas behind smart manufacturing have their roots in the computer science field. The foundations of smart manufacturing have been influenced by the principles mostly derived from the field of computers. While the manufacturing industry is expected to derive ongoing advantages from various concepts and emerging ideas, such as the potential disruptive impact of quantum computing, it possesses a distinct identity that may be encapsulated by six fundamental pillars. Smart manufacturing encompasses six fundamental pillars. The nomenclature and significance of these six pillars have undergone alterations; however, they have persisted in the realm of manufacturing throughout its historical trajectory.

Pillar 1: Advancements in Manufacturing Technology and Processes: It is anticipated that there will be a rise in manufacturing technologies and processes in the coming years. There will be the emergence of novel materials, components, and products (Kusiak, 2016). Additive manufacturing exemplifies a novel technological advancement that has stimulated the emergence of fresh materials, influenced product design and production processes, and facilitated the exploration of novel applications, including biomanufacturing. Specifically designed manufacturing equipment allows for the integration of varied operations; for example, machining centers can be machines that can handle drilling, horizontal and vertical milling, and so on. Emerging hybrid processes include the integration of traditional and additive processes, as well as the utilization of laser and net-shape production techniques. There will be an increased level of integration in several processes, such as the incorporation of novel materials, product design, and manufacturing procedures. For instance, the identification of a chemical component may prompt the development of a new medication and delivery method, as well as the subsequent production of both the drug and the device. The popularity of additive manufacturing in factories is expected to increase, including both large and small areas. The advent of affordable robots will significantly improve manufacturing automation. The integration of sensors and software capabilities will enhance the intelligence of the new manufacturing equipment, enabling seamless communication both within the facility and beyond.

Pillar 2: Materials: Shape memory alloys and functionally graded materials are examples of smart materials, but their development is not an explicit goal of smart manufacturing. Separate evolution paths for smart materials and smart products are not out of the question. Smart manufacturing is founded on the idea of utilizing multiple resources, including biomaterials and organic-based materials, to make future products easier to produce. There will be an increased priority placed on the recovery of materials from products at the conclusion of their lives. Landfills have the potential to transform into new sources of diverse resources (Kusiak, 2018). Certain novel materials necessitate the development and integration of innovative procedures within the realm of smart manufacturing. The utilization of additive manufacturing will significantly contribute to the exploration of novel materials and their combinations.

Pillar 3: Data: We are currently observing the revival of data in the manufacturing industry. The deployment of sensors, wireless technologies, and advancements in data analytics have contributed to the initiation of some phenomena. Data collection has started on a larger scale, encompassing a variety of sources such as material quality, process characteristics, customers, and suppliers. Everything from building prediction models to fueling any imagined application will make use of the data (Riordan et al., 2019). Furthermore, it will serve as the optimal resource for safeguarding and extracting historical and contemporary knowledge pertaining to manufacturing.

Pillar 4: Anticipatory Engineering: This is a recent advancement in the field of manufacturing solutions, aimed at fostering an anticipatory approach rather than a reactive one. In the past, the manufacturing sector has primarily prioritized the utilization of data for the purposes of analysis, monitoring, and control. Nevertheless, conventional endeavors have predominantly prioritized historical aspects rather than prospective developments in manufacturing methods and systems. The field of predictive engineering presents a novel approach to creating accurate models, namely digital representations, of the phenomena under investigation (Molina et al., 2021). These models will enable the exploration of future places, both inside the current technological framework and in previously unexplored areas. In forthcoming times, contemporary models will be enhanced with a combination of limited-scope models, such as those pertaining to the behavior of a supply chain, and models that encompass multiple systems. This integration aims to facilitate informed decision-making regarding future production and market conditions. Wide-ranging models have the potential to facilitate the restructuring of the manufacturing industry. It is plausible that certain industrial processes may have a highly distributed structure, while others may choose a centralized approach.

Pillar 5: Sustainability Over Time: The significance of sustainability in the manufacturing sector cannot be overstated. The objectives of sustainability initiatives encompass various aspects, including resources, manufacturing processes, energy consumption, and pollutants associated with manufacturing activities. Both the product and the market serve as the primary entrance points for any significant sustainability initiative (Romero and Stahre, 2021). Likely circumstances encompass the following: (1) the integration of sustainable product design into manufacturing, (2) the influence of sustainable manufacturing processes on product design, and (3) the simultaneous improvement of eco-friendly goods, procedures, and materials. Additive manufacturing is a process that has led to the creation of novel designs for components and products. Sustainability pertains to the manner in which a product is executed, rather than its actual production. Due to the principles of sustainability, the distinction between manufacturing and service will continue to be indistinct (Kaur et al., 2023). One illustration of a nontraditional manufacturing activity is the process of reconditioning a used product, which has the potential to be included in the new manufacturing vocabulary.

Pillar 6: Networking and Resource Sharing: With the increasing digitization and virtualization of production, a significant portion of creative and decision-making processes will occur in the digital realm. Although the digital world may possess a certain degree of transparency, the physical manufacturing assets and their expertise will remain safeguarded. The implementation of digital-physical separation facilitates the equitable utilization of resources among various enterprises, even those engaged in competition (Morris et al., 2020). Service and contract models have been implemented in the manufacturing business, where production occurs at facilities managed by a third party. The rapid manufacturing service model, which preceded 3D printing, was formed several decades ago due to the exorbitant cost of technology, limited adoption, steep learning curve, and uncertainty regarding the technology's usefulness. The popularity of shared resource models has led to their expansion beyond the initial purpose of decreasing highway traffic through ride-sharing services. This expansion is seen in the case of Uber in transportation and Airbnb in housing services. The principles of sharing manufacturing equipment, software, experience, and, notably, the collaborative modeling and creation space are expected to yield significant advantages for smart manufacturing (Kusiak, 2017).

5.2 IMPORTANCE OF SENSORS IN SMART MANUFACTURING

Sensors are utilized throughout many industries for a multitude of applications, encompassing both every day and commercial contexts. Sensor systems have recently gained prominence in industrial demonstrations, highlighting their remarkable capabilities. Sensors establish connections between several devices and systems, facilitating communication between different machines to monitor and trace systems. Sensors have undergone significant advancements in terms of their compactness and portability, enabling their integration into challenging-to-reach and potentially dangerous gadgets. This integration has transformed these devices into advanced technological tools. Sensors have been integral to manufacturing processes for an extended period (Javaid et al., 2021). However, until recently, challenges related to machine noise, signal attenuation, and reaction dynamics have predominantly been constrained. Currently, sensors are widely employed in daily activities. The notion spans a wide range of commodities, including those used in commercial, automotive, and military contexts. An autonomous system is characterized by its ability to make decisions on several variables. Due to their advanced onboard processing capabilities, sensors possess the ability to evaluate atmospheric conditions and subsequently modify operations accordingly (Konyha and Bányai, 2017). These judgments are made using extensive amounts of evidence and can be analyzed with greater speed and accuracy than anyone could anticipate. This mitigates all the hazards linked to human fallibility and guarantees that productivity and quality may be enhanced with minimal supervision. Quantitative analysis offers a notable benefit in terms of maintenance, as it can be efficiently performed utilizing this technology. Sensors facilitate the user's ability to accurately and remotely perceive and document tangible feedback from various sources, such as physical motion, fluctuations in temperature, and electrical signals. Intelligent connectors play a crucial role in guaranteeing the proper functioning of manufacturing applications by offering a comprehensive and forward-thinking device solution (Namjoshi and Rawat, 2022). Sensors play a crucial role in many industrial and commercial settings by providing vital insights, increasing production, promoting, selling, competing, reducing expenses, and improving corporate performance and overall productivity. Organizations strive to achieve self-optimized and interconnected production processes, with a primary focus on the processing and evaluation of data at lower levels within the production hierarchy.

5.2.1 What is a sensor?

The term "sensor" refers to any device that can take in information about the physical world—in the form of input stimuli—and then turn it into a digital signal with a quantifiable value. Pressure, flow, light, heat, motion, dampness, and a plethora of other environmental phenomena could all constitute the input stimuli. An electrical signal—voltage, current, capacitance, resistance, frequency, etc.—is a common format for the response's output. The next step is to either electrically transmit this signal over a network or convert it into a form that can be read by a human eye.

5.2.2 What is smart factory?

Manufacturers have been able to take factory automation to a new level with the help of AI and the IoT. Complex optimization decisions, previously performed by people, are now automated as a result of this combination of physical machinery and business processes. What we now call a "smart factory" was born out of manufacturers' ability to integrate floor decisions and perceptions with the supply chain. The advent of mechanical manufacturing equipment signified the initiation of the initial industrial revolution, which was subsequently succeeded by the advancement of mass production of commodities (Shrouf et al., 2014). The digital revolution is often regarded as the use of enhanced automation and control in manufacturing operations through the utilization of electronics and information technology. Deviation technology, made possible by the integration of IoT into various processes, enables machines and industries to reconfigure and optimize themselves. This allows them to modify their actions in reaction to shifts in instructions and operational circumstances (Dalenogare et al., 2018). The fundamental aspect of smart factories is in the technological infrastructure that enables the collecting of data.

5.2.3 The most important sensors for a smart factory

Intelligent machinery, gadgets, and control equipment are integral components of smart factories, which are responsible for monitoring critical parameters of industrial processes (Lass and Gronau, 2020). The aforementioned enhancements have not alone modified the infrastructure of the factory floor, facilitating consistent and accurate cooperation among machines, but have also modified the demands placed on machinery, resulting in an increased need for dependable sensors (Cimini et al., 2020). This section provides concise details regarding the primary sensors employed in a smart factory (Figure 5.1).

5.2.3.1 Passive sensors

Sensors play a crucial role in a smart factory by gathering and integrating precise data into manufacturing processes to improve product quality. They are specialized devices composed of sensitive materials that utilize electrical, opto-electrical, or electronic methods to detect the existence of a specific entity or function. In numerous instances, sensors are employed to convert

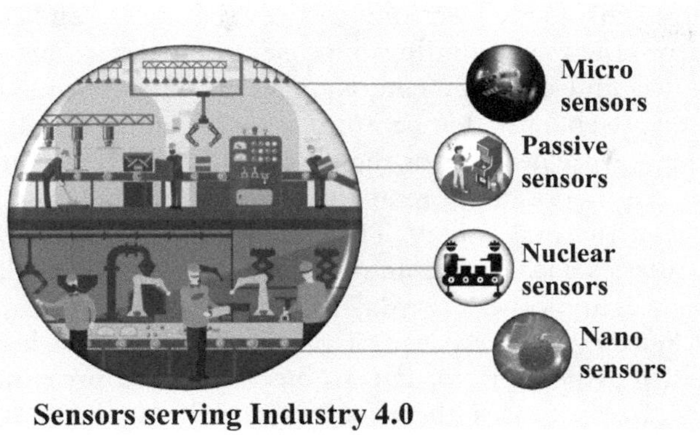

Figure 5.1 Various sensors used in industry.

a physical stimulus into an electrical signal, enabling subsequent evaluation and analysis to facilitate decision-making regarding the ongoing activities (Herrojo et al., 2019). Manufacturers now have unprecedented control and acquisition of data due to recent advancements in sensor technology. The operation of sensors can be classified as either active or passive for the sensor to function, it necessitates a specific physical stimulation during active operation. Color identification sensors are considered active due to their reliance on visible light for object illumination, which enables the sensor to detect a physical stimulation. The passive instance refers to a situation where the physical stimulus is already available and does not require any additional provision (Mullon et al., 2020). Numerous sensor variants have been devised and effectively employed in the realm of industrial process control. Intelligent manufacturing facilities employ a diverse range of sensor technologies, encompassing fundamental temperature and humidity monitoring, as well as advanced location and product sensing (Jeon et al., 2020). The utilization of these sensors enhances manufacturing efficiency by facilitating various industrial operations, including product transportation, robotic and milling process control, and environmental factor detection. In a production setting, the primary parameters for measurement and control encompass temperature, location, force, pressure, and flow (Landaluce et al., 2020). Sensors are categorized into different groups based on characteristics such as the type of measurement, areas where they are used, the mechanism of conversion, the energy range of the measurement, and thermodynamic concerns.

Pressure Sensor: Commonly, they are electromechanical devices that take readings of liquid or gas pressure and send those readings on to the control and display systems' inputs. Pressure sensors typically exhibit substantial variations in terms of their technology, design, applicability for certain applications, performance, and cost. These sensors are applied for the determination of controlling and intensive care a wide range of commonplace applications (Tessarolo et al., 2018).

Temperature Sensors are electronic devices that detect and convert thermal factors from their surroundings into electronic data. This data is then used to monitor and communicate changes in temperature. In order to detect temperature changes, temperature sensors commonly utilize either a resistance temperature detector or thermocouples. The thermometer is well recognized as the predominant temperature sensor employed for quantifying the extent of thermal energy. These sensors are commonly employed in several industries such as the automotive, aviation, medical, computer, kitchen appliance, and other everyday applications (Herter et al., 2018).

Proximity Sensor: With the use of proximity sensors, which are electronic devices, it is possible to detect objects in close proximity without touching them. In most cases, these sensors work by sending out an electromagnetic field and then picking up on changes to that field or a signal that corresponds to it. The presence of parts and machine components can be detected by these sensors in a variety of production processes. These devices are used to measure the variation in the distance between a shaft and its supporting bearing in machine vibration monitoring (Du et al., 2020).

Humidity Sensors: These sensors measure the amount of water in the air and convert those readings into signals that other devices can use as inputs. These ranges find widespread application in many fields, including industry, meteorology, medicine, automobiles, and heating, ventilation, and air conditioning. Sensors that detect pressure: One way to measure mechanical forces, such as compressive and tensile stresses, is with a force sensor, which is also called a load cell. You may learn about the applied force's magnitude from these signals. Among the many fields that make heavy use of these sensors are the ones dealing with consumer goods, computers, music, medicine, the automobile industry, and athletics (Chen et al., 2005).

Flow Sensors are sensors that measure the rate of change of a fluid (gas, liquid, etc.) and send signals back to a controller; these sensors can be electrical or electromechanical. The regulation of numerous industrial processes relies heavily on flow measurement, making these sensors essential in the medical and automotive industries (Lebossé et al., 2011).

Gas Sensors are electronic devices that can be either fixed or portable. They can identify the presence and properties of many different gases. After then, the sensors will send an output signal to the controller. Among other things, these sensors identify dangerous gases and provide a vital way to track gas concentration and environmental data, which solves a lot of safety problems.

Flaw Sensors: In order to detect and emphasize surface or underlying material defects, flaw sensors are widely used in many manufacturing processes. Detectors and sensors often use acoustic, ultrasonic, or other techniques to find and identify defects in different kinds of materials. Everyone agrees that these gadgets are must-haves for manufacturing quality control. Nondestructive testing and material analysis rely on their inspection of components (Petrov et al., 2021).

Color Sensor: A photoelectric sensor, like the color sensor, uses a transmitter to send out light beams and a receiver to pick up the reflected light from whatever it is trying to detect. Image processing, color identification, industrial item tracking, industrial process control, medical diagnosis systems, health fitness systems, and a host of other fields can all benefit from the development of color sensing applications made possible by these sensors.

Light Sensor: The light sensor, alternatively referred to as a photoelectric device, is responsible for the conversion of observed light energy (photons) into electrical energy (electrons). These instruments are extremely important in several industries. Nevertheless, several light sensors exhibit distinct operational characteristics and possess specialized applications (Gödecke et al., 2020).

5.2.3.2 Nuclear sensor

Similar to how optical sensors work—using a medium to transmit radiation from a source to a detector—the nuclear sensor does the same thing. The measured variable determines the reduction in the magnitude of this transmission. Medical imaging and mass flow measurement both make use of nuclear sensors.

5.2.3.3 Micro-sensors

In both two and three dimensions, micro-machined structures are microelectromechanical system devices that comprise micro-sensors. Because they convert mechanical impulses from an energy source into electrical signals, these sensors can be thought of as miniature transducers. A wide variety of sensors are already finding useful applications in industry, tracking variables such as heat, pressure, force, speed, sound, magnetic field, optics, biology, and chemistry (Shkel, 2001).

5.2.3.4 Nano-sensors

Nano-sensors, which are derived from nanotechnology, represent the latest advancements in the field of sensing technology. These devices are components of nano-electromechanical systems, which also encompass nano-actuators.

5.2.3.5 Smart sensors

Along with other noteworthy technical breakthroughs, smart sensors have recently attracted a lot of attention for their potential importance and wide range of application sectors. Traditional sensors have been upgraded to intelligent ones, capable of complex computations using collected data, thanks to the integration of computing and the IoT into manufacturing processes. Not only have smart sensors improved their functionality, but they have also become more smaller and more flexible, turning clumsy machines into high-tech marvels. With the addition of signal conditioning, integrated algorithms, and digital interfaces, smart sensors have progressed into devices with detection and self-awareness capabilities. The sensors are constructed as IoT components, which provide the conversion of real-time data into digital information, which may then be communicated to a gateway. These capabilities enable intelligent sensors to forecast and observe real-time situations and promptly implement remedial measures. The primary activities of intelligent sensors encompass intricate multilayered procedures, including the collection of raw data, and sensitivity adjustment. One example of an application of smart sensors is wireless sensor networks, wherein nodes are interconnected with one or more additional sensors and sensor hubs, thereby establishing a form of communication technology. Furthermore, the integration of data from several sensors can be employed to draw inferences regarding a preexisting issue (Ashima et al., 2021).

Capability for Calibration: The phrase "calibration capability" refers to the capacity of a sensor to ascertain its normal functioning. Self-calibration is often straightforward, and various calibration methods are accessible for different sensor types. For instance, load cells employed in weighing systems have the capability to adjust their output to zero in the absence of any applied force. Look-up tables can be utilized by other sensors for the purpose of calibration. In contrast, it is more advantageous to employ an interpolation technique when a limited number of correction points are necessary (Li et al., 2016).

Self-Identification of Defects: Intelligent sensors engage in self-diagnosis by monitoring internal signals to detect indications of malfunctions. Some sensors may face difficulties in distinguishing between typical measurement variations and sensor malfunctions. One approach to address this difficulty involves the storage of several measured values in proximity to a predetermined set-up point. Uncertainty techniques are employed to assess the influence of sensor malfunction on the measured amount. This facilitates the ongoing utilization of a sensor subsequent to the occurrence of a failure (Leng et al., 2019) (Figure 5.2).

5.3 NEED OF SENSORS IN INDUSTRY

The demand for sensors in the industrial sector arises from the necessity for a cohesive and intelligent network that facilitates intelligent production processes. Sensors have the capability to collect and analyze data in order to facilitate efficient decision-making, and it is feasible to optimize automation in production lines through self-optimization. Factory automation allows for the optimization of personalized product solutions and assembly lines.

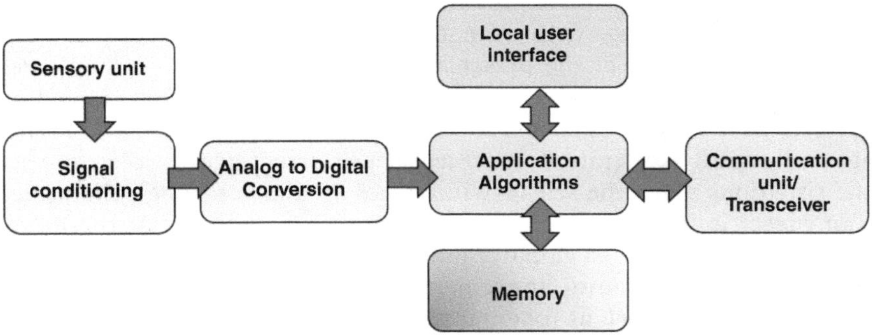

Figure 5.2 Need of sensors in industry.

In the forthcoming years, it is anticipated that the industry would adopt asset management technologies. Asset monitoring systems and technologies play a crucial role in the supply chain industry, facilitating efficient supply lines and driving significant demand in the sensor market.

5.4 USE OF SENSORS IN THE SMART MANUFACTURING SECTOR

Sensors are employed to gather data from the physical environment through the utilization of a transducer. It utilizes computer capacity to perform a predetermined and optimum action on the collected data. The data are subsequently transferred via a networked connection. The memory and architecture of intelligent sensors are constrained, which poses challenges in effectively eliminating noise and abnormalities from their outcomes. In the field of industry, computer communication plays a crucial role in facilitating the exchange of status and activity-related information among computers, as well as in the collection of supporting data and control orders to enhance the efficiency of their operations. Industry is implemented using sensor technologies (Kamble et al., 2018). The objective is to streamline the process of data collection by eliminating the need for laptops or tablets. Sensors are progressively incorporating automated manufacturing lines and using networking, IoT process control, data interchange, and logistical convergence. The production process will increasingly rely on long-range proximity sensors in varied shapes due to the ongoing transformation of industry into both vast and small areas. The sensor is typically equipped with communication capabilities and possesses a specific type of onboard diagnostic feature. Intelligent sensors offer a wide range of functions and possibilities. Smart sensors have the capability to do self-assays and auto-calibration (Zheng et al., 2018). In the future, sensors will be utilized to avoid injuries in athletes and aid in pain relief without the need for medicine. Companies employ these devices in settings with infrequent activity and, equipped with sensors, can readily detect the presence of individuals. These technologies are currently accessible on a vast number of computer systems, including smartphones. Their uses encompass vibration detection, inclination sensing, and overall acceleration. It serves as a deterrent against theft in specific situations when the sensor will trigger an alarm through the device if an object is in motion and needs to remain immobile. Furthermore, this technology has been employed in the context of the COVID-19 pandemic. Long-range sensor implementations in the industry encompass many applications such as manufacturing and robotics automation, agile production routing, intelligent production, retail inventory management, and supply chain management. The mechanism of self-optimization is significantly influenced by automation. The automated production systems are equipped with sensors positioned at their core. This comprehensive framework encompasses a wide range of applications, including the assembly of lines and conveyor production processes, as well as the prevention of robotic crashes and human–machine contacts. Robotic automation incorporates sensors to enhance the safety of robotic manual labor. These devices are specifically designed to examine ongoing work, monitor the connection between the work-in-progress and the production equipment, and enable self-monitoring of production using their production system computer. Sensor technologies are utilized in industrial facilities to detect, measure, analyze, and process various modifications, including position, weight, height, displacement, and presence (Abdolmaleki et al., 2021). The production business is currently undergoing a transformation in response to the potential benefits for producers, including reduced downtime, improved operating performance, and lower operating costs. Sensors possess a multitude of applications that perform a crucial function inside automated production systems.

5.4.1 To monitor the complete manufacturing process

In order to keep tabs on every step of production, businesses use sensors. These devices are utilized for the purpose of data collection and transmission to centralized cloud computing platforms for the purpose of industry analysis. The findings can still be monitored by the principal decision-makers. Intelligent sensors possess a high degree of versatility and find application across several industries. In the field of medicine, biological functions are quantified, including the assessment of blood flow during surgical procedures, as well as the monitoring of heat leaks in buildings and industrial plant structures during the stages of design, engineering, and construction (Zhong et al., 2017). In the retail industry, sensors are employed to identify the position of customers and monitor the movement of crowds.

5.4.2 The integration of different devices

Sensors are employed to enhance operational processes and optimize manufacturing efficiency, providing companies with an opportunity to transform their facilities. Intelligent sensors facilitate the generation of data through the interconnection of many devices and systems, enabling intercommunication among diverse units. By minimizing scheduled servicing, replacement expenses, and market disruption capacity, manufacturers are able to decrease the value of their substitute assets. Furthermore, intelligent technology plays a crucial role in enabling companies to shift from planned maintenance to automated maintenance. Data might reveal patterns, indicating the necessity for equipment maintenance (Pinto et al., 2021). Intelligent devices will utilize this data to provide users with alerts in order to prevent them from becoming potential sources of failure. It is anticipated that numerous vendors will furnish reports as evidence of their adherence.

5.4.3 To improve production performance

With the help of sensors, agile approaches can be easily implemented, leading to improved production performance in real-time processes. Industry will be able to provide visual representations of peaks and flows and increase plant transparency through the use of sensor data. The utilization of intelligent sensors and digitization in manufacturing allows firms to effectively manage their production processes in a transparent, dependable, and superior manner. Enhanced precision in the factory enables businesses to achieve more compliance and efficiency, hence enhancing production performance. The objective of intelligent production is to create a manufacturing facility that is digitally interconnected, enabling the integration of physical and cyber technology to enhance production efficiencies. Industry integrates distinct frameworks and leverages the potential of vast amounts of data to enhance automation.

5.4.4 Management of quality

Sensors can be used to track equipment platforms and aid with quality management in manufacturing settings. Innovations in sensor technology have attracted interest and made potential producers more effective. Intelligent sensors establish connections between several operating systems, enabling seamless communication and intelligent decision-making among different equipment. Microprocessors are employed to tailor outputs and offer interpretation, hence ensuring precise performance in comparison to traditional sensors. Consequently, factory managers are able to make informed judgments regarding their operations by utilizing precise data and facilitating effective communication amongst machines.

5.4.5 Precision in production machinery

The use of sensors in manufacturing machinery systems could lead to improvements in performance and accuracy. Automatic data recording is done by sensors for things like energy consumption, temperature, humidity, running time, maintenance, and outputs from the production line. In addition to mitigating regulatory enforcement pressure, intelligence sensors have the potential to enhance production processes. The system has the capability to identify anomalies in the device that could potentially hinder the performance and quality of the product, and promptly communicate these issues in real time. Furthermore, producers have the ability to promote economic advantages, broaden their range of services, optimize product variety, and enhance consumer loyalty (Borghetti et al., 2020).

5.4.6 To calculate the speed of machines

The speed of machines can be determined by the utilization of sensors. It is utilized for estimating angular and rotational speeds in three-dimensional directions. The primary purpose is to monitor the orientation of an object. The principal applications of this technology encompass several sectors such as automotive, gaming, and mobile devices. Data are generated in a constant manner by sensors for every facet of the production process, enabling real-time collection and analysis (Wijaya et al., 2017). The production operations management will thereafter acquire and generate this information in real-time.

5.4.7 To improve automation through sensing and communicating data

Sensors are devices that can passively gather data generated throughout various stages of manufacturing or logistics in order to improve automation. They facilitate the intelligent detection and communication of information by machines, allowing for the identification and analysis of data to enhance automation. The sensors' data allows for the operation of equipment and processes, as well as the tracking of specific production systems and the detection or prevention of potential failures. The incorporation of clarity into materials and processes will lead to their refinement. Currently, sensors play a crucial role in the automation and digitalization of factories (Mocnej et al., 2016). It provides data for the purpose of retrieving specific information without engaging in computational processes. These systems collect diverse inputs and compute several parameters through the utilization of multiple sensor kinds. Various types of proximity sensors, such as inductive, photoelectrical, or magnetic sensors, can be utilized to detect movement.

5.4.8 Process regulation

To ensure the ongoing monitoring and control of processes, a multitude of sensors assume crucial functions. IoT-enabled enterprises are equipped with numerous sensors that significantly contribute to industrial control by virtue of their remote sensing and tracking capabilities. In the realm of tailored automation, product development solutions are employed to monitor the movement of materials. Sensors play a crucial role in contemporary manufacturing lines across several industries (Wang et al., 2021). Traditional conveyor manufacturers incorporate long-distance sensors into their systems in order to enhance production automation capabilities and minimize disruptions caused by proximity.

5.4.9 Fault detection

Through the continuous integration of industrial sensors into equipment, predictive analytics may detect possible issues and determine how long specific parts are at their most efficient. In order to minimize possible delays, factory managers can use the information provided to make sure that new components are on-site and to strategically schedule for repairs. Nevertheless, human error continues to be a significant contributor to device failure. This phenomenon can be attributed to the consideration of several aspects, such as the presence of competent personnel while ensuring optimal efficiency and safety within contemporary manufacturing settings (Sahoo and Samal, 2023).

5.4.10 Data transfer

Sensors are specifically engineered to collect and convey data to a nearby network. These sensors can vary in size and can be used on many surfaces and applications, unlike traditional sensors that can only determine the accessibility of an item. Intelligent sensors have the capability to collect a substantial volume of data. These devices have the ability to detect temperature, moisture, stress, and other factors based on their arrangement. Sensors can be integrated into an IoT network to oversee the temperature of drugs, the atmosphere for delicate electronics, and the growth of microorganisms in food. Assembly lines that handle goods with varying dimensions or forms can resume production operations without experiencing any deceleration through the utilization of sensors (Prato et al., 2019).

5.4.11 Pressure measurement

An electrical signal is converted into pressure using a pressure sensor. In this particular scenario, the quantity is contingent upon the level of pressure. Numerous systems rely on liquid or alternative sources of pressure.

The utilization of these sensors enables the development of IoT systems for monitoring pressurized systems. Water consistency sensors are mostly employed by water distributors to assess water quality and identify ions (Choudhary et al., 2014). Nevertheless, these sensors serve a crucial function in the surveillance of water quality for diverse objectives. Consequently, it finds use across a diverse array of industries.

5.4.12 To enhance production efficiency

Sensor-based job monitoring enables the reduction of idle working hours by optimizing production operations. To effectively address potential disruptions, sensors are employed to connect and monitor key infrastructure. In the context of Industry, the implementation of automation in plants enables the autonomous optimization of production lines and the provision of tailored solutions (Dutta et al., 2020). Automation is progressively utilizing a range of sensors, including temperature, pressure, and torque sensors.

5.4.13 To identify underlying deficiencies in the production process

The implementation of sensors leads to enhanced operational efficiencies, including reduced costs in workforce, transportation, and quality management. Additionally, they assist in identifying underlying flaws in production while enhancing product design. As an illustration, wearable assembly lines have the capability to transmit real-time photos to design engineers by means of clever sensor technologies. The implementation of stock counts, material sorting, and automation has resulted in enhanced efficiency for sensors (Haleem et al., 2021). Sensors can facilitate the generation of reports in data-equipped equipment to demonstrate adherence to industry regulatory standards. In addition, sensors frequently assume a significant role in enhancing manufacturing processes through the provision of product quality performance and the timely reporting of issues.

5.5 LOGISTICS TRANSPORTATION OF RESOURCES, COMPONENTS, PRODUCTS, AND INDIVIDUALS

The functional areas of material handling and transport in manufacturing encompass distances that range from nanometers to kilometers. Both facilitate the manufacturing operations and transportation of goods between various factories and worldwide. Material handling typically refers to the process of moving items within an industrial setting. The cost of a product can be significantly influenced by material handling and transportation. For instance, Cotrell et al. (2014) found that 8% of the cost of a wind turbine tower is ascribed to transportation, while a higher percentage (e.g., 20%) is attributed to other components. The spread nature of manufacturing is expected to result in transportation of materials, components, products, and people being a substantial cost in production. The implementation of this approach will inherently result in the optimization of transportation and utilization expenses associated with staff who provide support for the physical and digital infrastructure across various manufacturing plants (Kusiak, 2018). Job descriptions that are currently nonexistent will be developed to address the requirements of smart manufacturing tasks. Manufacturing specialists, like components, may engage in extensive travel using various means of transportation, such as cars, trains, and plains, to differentiate their products. As further elaborated upon in this scholarly article, akin to the allocation of manufacturing resources, it is probable that human resources will be distributed in substantial quantities. The manufacturing cost is influenced by the efficiency of transferring resources, components, products, and people. Transport and communication play a pivotal role in facilitating global connectivity within the manufacturing sector. Regardless of ownership, transportation is expected to play a crucial role in smart manufacturing as a result of several factors: (1) increased dependence on the transportation of materials, components, products, and service workers to meet individual requirements; (2) the importance of sustainability; and (3) the necessity for high-quality service. Transport networks that encompass both the supply side and distribution, including client delivery, as seen in Figure 5.3, are expected to have a significant impact. Restricting the examination of sustainability solely to the manufacturing envelope would yield a less-than-ideal resolution, since it fails to account for the interdependence between manufacturing processes and the trade-offs associated with the

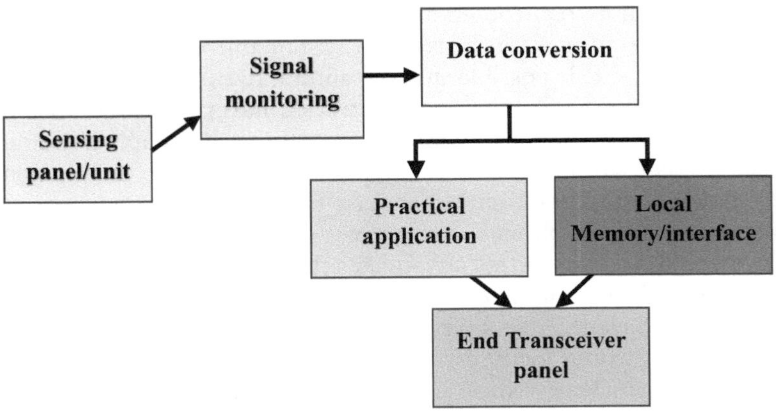

Figure 5.3 Step-by-step working of sensors.

supply and distribution networks. The correlation between the quality of customer service and factors such as inventory level, production response time, and transportation are significant.

5.5.1 Smart automobiles

Currently, numerous material handling and transportation vehicles are engaged in the transmission and reception of data. The level of vehicle connectivity will increase, leading to enhanced involvement in the flow of information, such as communication between vehicles or between vehicles and maintenance centers for remote diagnosis and repair. Indeed, it is probable that machine tools will adhere to a similar trajectory of communication that facilitates condition monitoring and production. All vehicles, regardless of their fuel type and automation/use attribute, are included. As an illustration, it is evident that a car has the potential to be both autonomous and electric (Oyekanlu et al., 2020). This pertains to material handling systems, private automobiles, recreational vehicles, and public transportation. There seems to be a natural inclination within technology to establish connections between transportation, energy, sustainability, and manufacturing. Various types of vehicles, such as forklifts, cars, trucks, and long-distance trains, have the potential to be electric and autonomous. Furthermore, cars have the potential to be shared. Currently, electric car batteries are charged using electricity typically produced by conventional power plants. In forthcoming times, automobiles will employ electricity derived from renewable sources, such as hydroelectric power provided by wind turbines or hydrogen, and will be constructed with a focus on sustainability, encompassing energy efficiency. A vehicle that is energy-efficient will utilize a less amount of material and will be designed in a manner that minimizes aerodynamic drag (Mehami et al., 2018). Upon reaching the end of its lifespan, the majority of its constituent parts will be repurposed and recrafted. Thus far, there has been a dearth of scholarly focus on the topic of sustainable design in the context of electric vehicles. It is possible that the automotive industry is excessively focused on achieving a prosperous penetration into the autonomous market. It is not a novel concept to consider vehicle connectivity. The concept has been present within the realm of public transportation for an extended duration. The implementation in the mass-transit arena encompassed two distinct types, namely physical connectivity and information connectivity. Traditional connectivity offers advantages that can be extended to the latest generation of cars and material handling technologies. Vehicles, akin to machine tools, have the capability to establish physical or virtual connections. There are various methods available to establish a virtual connection, encompassing vehicle-to-vehicle communication as well as traffic control (Al-Turjman et al., 2022). For instance, akin to the manufacturing industry where a single operator may oversee multiple machines, the utilization of a single driver per numerous vehicles presents a cost-effective benefit. In the realm of personal vehicles, the concept of shared transportation has been in existence for a considerable period of time. Typically, the individual who has the car is also among the occupants. The practice of shared group transportation is well recognized and established, encompassing various scales such as small groups (e.g., six to ten individuals in a van) to large-scale mass transit systems. Utilizing a privately owned vehicle for personal transportation is deemed wasteful in terms of energy consumption and financial expenditure. Personal automobile usage is typically minimal. Nevertheless, we often overlook the fact that our true requirement is a transportation service, without the requirement of having a vehicle. The use of the transportation service model provides enhanced vehicle utilization and decreased transportation expenses. The heightened use subsequently decreases the quantity of automobiles on the roadways, so exerting a beneficial influence on traffic congestion and the environment. The batteries are experiencing an increase in capacity (Qu et al., 2019). Currently, it is enough for the majority of journeys undertaken by car. Prominent automotive manufacturers, such as Ford, BMW, and Volvo, have made commitments to introduce autonomous vehicles, potentially lacking steering wheels and pedals, within the forthcoming years. Regardless of the degree of autonomy, it is anticipated that future vehicles would exhibit increased automation, connectivity, energy efficiency, and enhanced safety feature (Tao et al., 2018). The objective is that automobiles and factories will be powered by electricity derived from renewable sources. The manufacturing facilities will employ sustainable materials in order to produce components that align with the transportation requirements. The transition from conventional transportation to autonomous systems does not possess a singular, definitive switch or timeframe. There is a lack of clearly defined objectives regarding the degree of autonomy. Similar to every emerging technology, the concept of transport autonomy will undergo a steady progression, first with specific applications in narrow sectors and subsequently extending to more appropriate domains, such as manufacturing supply and distribution networks (Mittal et al., 2019). The distinction between personal and mass transportation modes may not be readily apparent in the context of vehicle-sharing. It is evident that transportation is more cost-effective when goods and passenger loads are consolidated, such as through the use of two interconnected trucks, as opposed to employing several vehicles with restricted load capacity in each vehicle. In addition to their autonomy, vehicles are increasingly enhancing their reliability. Predictive engineering solutions aim to forecast forthcoming occurrences, encompassing the possibility of a vehicle component requiring maintenance or a probable collision. The decision system integrated within the system will provide optimal choices for action to mitigate the risk of component failure or accidents that could potentially disrupt the supply schedule (Ziebinski et al., 2021).

5.6 CHALLENGES AND OPPORTUNITIES IN MANUFACTURING

5.6.1 Challenges

With the advancement of intelligent manufacturing systems, there has been a notable rise in the utilization of machines to perform a greater number of processes. It is evident that several machines possess the capability to perform either identical or distinct activities or jobs. The strength of these connections may also exhibit dynamic variations based on the specific tasks being performed. Essentially, the collaboration between different machines has become crucial for the overall efficiency of current and future systems. The Internet of Machines (IoMT) utilizes a diverse range of sensors to consistently observe and assess the state of machines. Machine signatures, which are sensor outputs, offer an exceptional chance for making optimal decisions in manufacturing (Arents and Greitans, 2022). However, in order to fully harness the promise IoMT for smart manufacturing, it is crucial to address the following obstacles.

1. The initial obstacle is in determining the current condition of each machine. This status encompasses not only the state of being occupied or unoccupied, but also the state of one's health, specifically in terms of its proper functioning. The significance of this status information is in its ability to ascertain the reliability of this equipment in carrying out tasks. The most direct approach is to utilize sensors capable of performing both the sensing function and doing analysis through signal processing of the sensed data. The power source for these sensors might be either a connected supply or batteries. Nevertheless, due to the growing quantity of machines and their projected lifespan of one to two decades, there are situations when it becomes challenging to supply wired power or battery assistance. Wires restrict the mobility of sensors, as an illustration. The process of replacing batteries for various sensor systems can be both hard and time-consuming. Moreover, batteries may exhibit compromised safety and efficiency in some harsh environmental conditions.
2. The second problem pertains to the utilization of machine statuses in order to allocate work to individual units. Is it necessary for each machine to adhere to a rigid static schedule? The possibility of machine malfunctions, as well as the dynamic fluctuations in system-level tasks, can lead to schedules that are optimized differently in terms of energy consumption and overall operating time, among other factors. In such situations, it is crucial to allocate work to machines in a dynamic manner, taking into account the perceived condition of each unit.
3. The emergence of the third problem can be attributed to the intercommunication among machines, perhaps encompassing the presence of coordinating machines. Currently, certain jobs are performed by

machines originating from many sources, including different countries. This highlights the potential for issues in the dependability of communication and its influence on the cooperation among these devices. Furthermore, manufacturing assets are inherently closed systems that cannot be completely regulated externally, despite the presence of a bidirectional information exchange. Consider a machine tool as an illustration. G code can be transmitted to the machine for execution; however, direct control over the servos and spindles of the machines is not possible. This presents an additional significant obstacle that needs to be surmounted in order to facilitate complete control and automation.

5.6.2 Opportunities

Upgrade Outdated Machinery to Enable Intelligent Manufacturing: Despite the emergence of novel manufacturing technology and emerging enterprises, numerous established manufacturing organizations are lagging behind in adapting to the rapid digital transformation. The presence of legacy machinery in several small manufacturing enterprises is a frequent occurrence. Legacy machines, however, are highly beneficial for manufacturing organizations and extensively used in production. Consequently, small manufacturers are seeing a decline in their competitive edge within the global market due to their restricted access to information and their inadequate capacity to navigate the heightened intricacies of contemporary manufacturing landscapes (Baldea et al., 2017). The establishment of IoT connectivity between legacy machines is a fundamental aspect of the challenge.

Status Detection of Self-Powered Machines: It is crucial to ascertain the condition of a machine equipped with an autonomous power supply mechanism. By implementing self-powered sensors, the requirement for a wireline connection or battery to supply electricity is eliminated. By including supplementary wireless pairing and data transmission capabilities, this sensing system has the potential to be effectively employed in numerous machines, hence improving portability and mitigating maintenance expenses.

The Cooperation of IoMT Devices: Optimizing the synergy among a group of remotely collaborating devices is of significant importance. Communication between devices in distant locations may experience a momentary interruption due to physical distance and poor message methods. When a machine encounters isolation from other machines, what actions should it undertake? What is the optimal design for the central coordinator?

5.7 FUTURE PROSPECTS

It is anticipated that smart sensors will become more widely available in near future. Furthermore, these sensors and diagnostic tools, along with other linked devices, will be able to communicate with one another more easily thanks to wireless technology. Manufacturing will increasingly rely on sensors in their regular operations. On the outskirts of conventional cloud computing networks, sensors will provide the gadget with the resources it needs to operate. Not only that, it makes it easier for devices to be operated remotely, which means they may collect data and do basic computations before being linked back to a main network. There is a close relationship between the enterprise's main servers and sensor-equipped industrial equipment. Predictive maintenance on engines and pumps will be the primary use case for the sensor. Through the constant collecting of data, operators can spot irregularities in their early phases. It functions in the context of wireless networks. As a result of its appropriateness for repair people with restricted alternatives, it will collect data from another planet. With the use of sensors, operational processes can be better seen, and data can be collected and analyzed to optimize performance with no interruption, both of which increase productivity. When it comes to maintaining the future, sensors are vital. The term "sensor" refers to small, wireless devices that can scan and communicate, particularly designed to pick up on data on plant pressure, surface temperature, and vibrations.

REFERENCES

Abdolmaleki, H., Kidmose, P., & Agarwala, S. (2021). Droplet-based techniques for printing of functional inks for flexible physical sensors. *Advanced Materials*, *33*(20), 2006792.

Al-Turjman, F., Zahmatkesh, H., & Shahroze, R. (2022). An overview of security and privacy in smart cities' IoT communications. *Transactions on Emerging Telecommunications Technologies*, *33*(3), e3677.

Arents, J., & Greitans, M. (2022). Smart industrial robot control trends, challenges and opportunities within manufacturing. *Applied Sciences*, *12*(2), 937.

Ashima, R., Haleem, A., Bahl, S., Javaid, M., Mahla, S. K., & Singh, S. (2021). Automation and manufacturing of smart materials in additive manufacturing technologies using internet of things towards the adoption of industry 4.0. *Materials Today: Proceedings*, *45*, 5081–5088.

Baldea, M., Edgar, T. F., Stanley, B. L., & Kiss, A. A. (2017). Modular manufacturing processes: Status, challenges, and opportunities. *AIChE Journal*, *63*(10), 4262–4272.

Borghetti, M., Cantù, E., Sardini, E., & Serpelloni, M. (2020). Future sensors for smart objects by printing technologies in Industry 4.0 scenario. *Energies*, *13*(22), 5916

Chen, X., Vierling, L., & Deering, D. (2005). A simple and effective radiometric correction method to improve landscape change detection across sensors and across time. *Remote Sensing of Environment*, *98*(1), 63–79.

Choudhary, G., Jankowich, M., & Wu, W. C. (2014). Elevated pulmonary artery systolic pressure predicts heart failure admissions in African Americans: Jackson Heart Study. *Circulation: Heart Failure*, *7*(4), 558–564.

Cimini, C., Pirola, F., Pinto, R., & Cavalieri, S. (2020). A human-in-the-loop manufacturing control architecture for the next generation of production systems. *Journal of Manufacturing Systems*, *54*, 258–271.

Cotrell, J., Stehly, T., Johnson, J., Roberts, J. O., Parker, Z., Scott, G., & Heimiller, D. (2014). *Analysis of Transportation and Logistics Challenges Affecting the Deployment of Larger Wind Turbines: Summary of Results* (No. NREL/TP-5000-61063). National Renewable Energy Lab. (NREL), Golden, CO (United States).

Dalenogare, L. S., Benitez, G. B., Ayala, N. F., & Frank, A. G. (2018). The expected contribution of Industry 4.0 technologies for industrial performance. *International Journal of Production Economics*, *204*, 383–394.

Du, R. H., Liang, L. R., Yang, C. Q., Wang, W., Cao, T. Z., Li, M., & Shi, H. Z. (2020). Predictors of mortality for patients with COVID-19 pneumonia caused by SARS-CoV-2: A prospective cohort study. *European Respiratory Journal*, *55*(5).

Dutta, G., Kumar, R., Sindhwani, R., & Singh, R. K. (2020). Digital transformation priorities of India's discrete manufacturing SMEs–a conceptual study in perspective of Industry 4.0. *Competitiveness Review: An International Business Journal*, *30*(3), 289–314.

Edgar, T. F., & Pistikopoulos, E. N. (2018). Smart manufacturing and energy systems. *Computers & Chemical Engineering*, *114*, 130–144.

Fathi-moghaddam, M., Sadrabadi, M. T., & Rahmanshahi, M. (2018). Numerical simulation of the hydraulic performance of triangular and trapezoidal gabion weirs in free flow condition. *Flow Measurement and Instrumentation*, *62*, 93–104.

Gödecke, M. L., Bett, C. M., Buchta, D., Frenner, K., & Osten, W. (2020). Optical sensor design for fast and process-robust position measurements on small diffraction gratings. *Optics and Lasers in Engineering*, *134*, 106267.

Groumpos, P. P. (1995). The challenge of intelligent manufacturing systems (IMS): The European IMS information event. *Journal of Intelligent Manufacturing*, *6*, 67–77.

Haleem, A., Javaid, M., Singh, R. P., & Suman, R. (2021). Telemedicine for healthcare: Capabilities, features, barriers, and applications. *Sensors International*, 2, 100117.

Herrojo, C., Paredes, F., Mata-Contreras, J., & Martín, F. (2019). Chipless-RFID: A review and recent developments. *Sensors*, *19*(15), 3385.

Herter, T. L., Adams, J. D., Gull, G. E., Schoenwald, J., Keller, L. D., Pirger, B. E., & Ennico, K. (2018). FORCAST: A mid-infrared camera for SOFIA. *Journal of Astronomical Instrumentation*, 7(4), 1840005.

Javaid, M., Haleem, A., Singh, R. P., Khan, S., & Suman, R. (2021). Blockchain technology applications for Industry 4.0: A literature-based review. *Blockchain: Research and Applications*, 2(4), 100027.

Jeon, S., Ko, M., Lee, J., Choi, I., Byun, S. Y., Park, S., ... & Kim, S. (2020). Identification of antiviral drug candidates against SARS-CoV-2 from FDA-approved drugs. *Antimicrobial Agents and Chemotherapy*, *64*(7), 10–1128.

Kamble, S. S., Gunasekaran, A., & Gawankar, S. A. (2018). Sustainable Industry 4.0 framework: A systematic literature review identifying the current trends and future perspectives. *Process Safety and Environmental Protection*, *117*, 408–425.

Kaur, R., Gabrijelčič, D., & Klobučar, T. (2023). Artificial intelligence for cybersecurity: Literature review and future research directions. *Information Fusion*, *97*, 101804.

Konyha, J., & Bányai, T. (2017). Sensor networks for smart manufacturing processes. *Solid State Phenomena*, *261*, 456–462.

Kusiak, A. (1990). *Intelligent Manufacturing System*. Prentice-Hall, Englewood Cliffs, NJ.

Kusiak, A. (2016). Renewables: Share data on wind energy. *Nature*, *529*(7584), 19–21.

Kusiak, A. (2017). Smart manufacturing must embrace big data. *Nature*, *544*(7648), 23–25.

Kusiak, A. (2018). Smart manufacturing. *International Journal of Production Research*, *56*(1–2), 508–517.

Landaluce, H., Arjona, L., Perallos, A., Falcone, F., Angulo, I., & Muralter, F. (2020). A review of IoT sensing applications and challenges using RFID and wireless sensor networks. *Sensors*, *20*(9), 2495.

Lass, S., & Gronau, N. (2020). A factory operating system for extending existing factories to Industry 4.0. *Computers in Industry*, *115*, 103128.

Lebossé, C., Renaud, P., Bayle, B., & de Mathelin, M. (2011). Modeling and evaluation of low-cost force sensors. *IEEE Transactions on Robotics*, *27*(4), 815–822.

Leng, J., Jiang, P., Xu, K., Liu, Q., Zhao, J. L., Bian, Y., & Shi, R. (2019). Makerchain: A blockchain with chemical signature for self-organizing process in social manufacturing. *Journal of Cleaner Production*, *234*, 767–778.

Li, J., Galley, M., Brockett, C., Spithourakis, G. P., Gao, J., & Dolan, B. (2016). A persona-based neural conversation model. arXiv preprint arXiv:1603.06155.

Mehami, J., Nawi, M., & Zhong, R. Y. (2018). Smart automated guided vehicles for manufacturing in the context of Industry 4.0. *Procedia Manufacturing*, *26*, 1077–1086.

Mittal, S., Khan, M. A., Romero, D., & Wuest, T. (2019). Building blocks for adopting smart manufacturing. *Procedia Manufacturing*, *34*, 978–985.

Mocnej, J., Lojka, T., & Zolotová, I. (2016, January). Using information entropy in smart sensors for decentralized data acquisition architecture. In: *2016 IEEE 14th International Symposium on Applied Machine Intelligence and Informatics (SAMI)* (pp. 47–50). IEEE, Herlany.

Molina, M. D., Sundar, S. S., Le, T., & Lee, D. (2021). "Fake news" is not simply false information: A concept explication and taxonomy of online content. *American Behavioral Scientist*, *65*(2), 180–212.

Morris, C., Kriege, N. M., Bause, F., Kersting, K., Mutzel, P., & Neumann, M. (2020). Tudataset: A collection of benchmark datasets for learning with graphs. arXiv preprint arXiv:2007.08663.

Mullon, C. D., Wakano, J. Y., & Ohtsuki, H. (2020). Evolutionary feedbacks between genes and their extended effects under limited gene flow and biased ecological inheritance. *BioRxiv*.

Namjoshi, J., & Rawat, M. (2022). Role of smart manufacturing in industry 4.0. *Materials Today: Proceedings*, *63*, 475–478.

Oyekanlu, E. A., Smith, A. C., Thomas, W. P., Mulroy, G., Hitesh, D., Ramsey, M.,& Sun, D. (2020). A review of recent advances in automated guided vehicle technologies: Integration challenges and research areas for 5G-based smart manufacturing applications. *IEEE Access*, *8*, 202312–202353.

Pereira, T., Barreto, L., & Amaral, A. (2017). Network and information security challenges within Industry 4.0 paradigm. *Procedia Manufacturing*, *13*, 1253–1260.

Petrov, V. V., Sysoev, V. V., Starnikova, A. P., Volkova, M. G., Kalazhokov, Z. K., Storozhenko, V. Y., & Bayan, E. M. (2021). Synthesis, characterization and gas sensing study of ZnO-SnO2 nanocomposite thin films. *Chemosensors*, *9*(6), 124.

Pinto, C. A., Farinha, J. T., & Singh, S. T. (2021). Contributions of petri nets to the reliability and availability of an electrical power system in a big European hospital: A case study. *WSEAS Transactions on Systems and Control*, *16*(2), 21–42.

Prato, L., Lindley, L., Boyles, M., Robinson, L., & Abley, C. (2019). Empowerment, environment and person-centred care: A qualitative study exploring the hospital experience for adults with cognitive impairment. *Dementia*, *18*(7–8), 2710–2730.

Qu, M., Nida, A., Kong, Y., Du, H., Xiao, G., & Wang, D. (2019a). Nanopolystyrene at predicted environmental concentration enhances microcystin-LR toxicity by inducing intestinal damage in Caenorhabditis elegans. *Ecotoxicology and Environmental Safety*, *183*, 109568.

Qu, Q., Zhai, Y., Li, X., Zhang, Y., & Zhu, Z. (2019b, September). Geometric analysis of nonconvex optimization landscapes for overcomplete learning. In: *International Conference on Learning Representations*. New Orleans, LA.

Riordan, B., Flor, M., & Pugh, R. (2019, August). How to account for mispellings: Quantifying the benefit of character representations in neural content scoring models. In: *Proceedings of the Fourteenth Workshop on Innovative Use of NLP for Building Educational Applications* (pp. 116–126). Florence.

Romero, D., & Stahre, J. (2021). Towards the resilient operator 5.0: The future of work in smart resilient manufacturing systems. *Procedia Cirp*, *104*, 1089–1094.

Sahoo, A. K., & Samal, S. K. (2023). Online fault detection and classification of 3-phase long transmission line using machine learning model. *Multiscale and Multidisciplinary Modeling, Experiments and Design*, *6*(1), 135–146.

Sheth, A., & Kusiak, A. (2022). Resiliency of smart manufacturing enterprises via information integration. *Journal of Industrial Information Integration*, *28*, 100370.

Shkel, A. M. (2001, August). Micromachined gyroscopes: Challenges, design solutions, and opportunities. In: V. K. Varadan (ed.), *Smart Structures and Materials 2001: Smart Electronics and MEMS* (Vol. 4334, pp. 74–85). SPIE.

Shrouf, F., Ordieres, J., & Miragliotta, G. (2014, December). Smart factories in Industry 4.0: A review of the concept and of energy management approached in production based on the Internet of Things paradigm. In: *2014 IEEE International Conference on Industrial Engineering and Engineering Management* (pp. 697–701). IEEE, Selangor Darul Ehsan.

Tao, F., Zhang, H., Liu, A., & Nee, A. Y. (2018). Digital twin in industry: State-of-the-art. *IEEE Transactions on Industrial Informatics*, *15*(4), 2405–2415.

Tessarolo, M., Gualandi, I., & Fraboni, B. (2018). Recent progress in wearable fully textile chemical sensors. *Advanced Materials Technologies*, *3*(10), 1700310.

Wang, M., Huang, W. J., Kondev, F. G., Audi, G., & Naimi, S. (2021). The AME 2020 atomic mass evaluation (II). Tables, graphs and references. *Chinese Physics C*, *45*(3), 030003.

Wijaya, W., Patel, A. R., Setiowati, A. D., & Van der Meeren, P. (2017). Functional colloids from proteins and polysaccharides for food applications. *Trends in Food Science & Technology*, *68*, 56–69.

Yao, X., Zhou, J., Lin, Y., Li, Y., Yu, H., & Liu, Y. (2019). Smart manufacturing based on cyber-physical systems and beyond. *Journal of Intelligent Manufacturing*, *30*, 2805–2817.

Zheng, Z., Xie, S., Dai, H. N., Chen, X., & Wang, H. (2018). Blockchain challenges and opportunities: A survey. *International Journal of Web and Grid Services*, *14*(4), 352–375.

Zhong, R. Y., Peng, Y., Xue, F., Fang, J., Zou, W., Luo, H., ... & Huang, G. Q. (2017). Prefabricated construction enabled by the internet-of-things. *Automation in Construction*, *76*, 59–70.

Ziebinski, A., Mrozek, D., Cupek, R., Grzechca, D., Fojcik, M., Drewniak, M., & Biernacki, P. (2021, June). Challenges associated with sensors and data fusion for AGV-driven smart manufacturing. In: *International Conference on Computational Science* (pp. 595–608). Springer, Cham.

Chapter 6

Smart sensors enabling navigation and mapping for smart farming

Anita Jaswal, Maninder Singh, and Prantick Patra

6.1 INTRODUCTION TO SMART SENSORS

Smart sensors are innovative devices that integrate conventional sensor characteristics with sophisticated computing capabilities, thereby facilitating the intelligent collection, processing, and transmission of data. Smart sensors, unlike traditional sensors, integrate embedded processors, memory, and communication interfaces, enabling them to independently carry out complicated tasks. Smart sensors are fundamentally engineered to optimize efficiency, precision, and responsiveness in a wide range of applications spanning many sectors (Spencer et al., 2004). Through the integration of sensing components and computational capabilities, these sensors possess the ability to locally evaluate data, derive significant insights, and transmit actionable information to other systems or people. An essential characteristic of intelligent sensors is their capacity to adjust and enhance operations in response to fluctuating environmental circumstances or user demands (Sehrawat and Gill, 2019). They possess the capability to modify sampling rates, selectively process data, and promptly activate alarms or warnings, making them indispensable for monitoring and control applications. In addition, intelligent sensors play a crucial role in Internet of Things (IoT) framework, facilitating the smooth incorporation of tangible items into interconnected networks. They function as the primary data collectors in IoT implementations, enabling decision-making based on data, anticipatory maintenance, and intelligent automation. Smart sensors fundamentally signify a fundamental change in sensor technology, providing enterprises and consumers with improved capacities to gather, analyze, and communicate data. The ongoing evolution of technology has positioned smart sensors to assume a crucial role in determining the future of several sectors such as smart cities, healthcare, transportation, manufacturing, and others.

DOI: 10.1201/9781003633884-6

6.1.1 Overview of smart sensor technologies

Smart sensor technologies are of paramount importance in the domains of localization, navigation, and mapping, as they offer instantaneous data pertaining to the surrounding environment. Below is a summary of how various technologies are employed in this particular situation: global navigation satellite systems (GNSS) such as global positioning system (GPS), GLONASS, Galileo, and BeiDou, employ satellite signals to ascertain the precise location, speed, and temporal data of terrestrial receivers. Accurate outside localization and navigation in autos, cellphones, drones, and maritime vessels heavily rely on these systems. Inertial measurement units (IMUs) comprise accelerometers, gyroscopes, and occasionally magnetometers, which are utilized to quantify the linear and angular movement of an object (Zhang et al., 2004). Continuous localization and navigation in regions with poor or unavailable satellite signals, such as indoor environments or urban canyons, are frequently achieved by the integration of these devices with GNSS systems. Light detection and ranging (LiDAR) is a technology that utilizes laser pulses to emit and measure the duration it takes for these pulses to return after colliding with objects. This process enables the creation of intricate three-dimensional (3D) maps of the surrounding environment (Indri et al., 2019). LiDAR technology is extensively employed in the domains of autonomous vehicles, robotics, and mapping applications to achieve precise localization and navigation within intricate surroundings. Visual odometry methods employ cameras to determine the motion of a system by tracking visual elements in consecutive photos. Simultaneous localization and mapping (SLAM) algorithms integrate input from several sensors, including cameras, LiDAR, and IMUs, in order to concurrently generate maps of an unfamiliar area and determine geographical coordinates within it. Autonomous robotics, drones, and augmented reality applications heavily rely on these techniques. Radio-frequency identification (RFID) technology is employed for the purposes of indoor localization and asset tracking, utilizing RFID tags and readers. RFID systems have the capability to detect and monitor things that are linked to RFID tags, allowing for various uses in logistics, inventory control, and interior navigation (Alonso et al., 2020). Ultrasonic sensors are utilized for the purpose of emitting and receiving ultrasonic waves in order to accurately determine the distances to various objects. Indoor localization systems often employ them, particularly in settings where GPS signals exhibit unreliability. The utilization of ultrasonic sensors in robotics and smart buildings enables the acquisition of accurate distance measurements, facilitating obstacle avoidance and localization. Magnetometers are utilized to quantify the magnetic field of the Earth and are employed in conjunction with other sensors to estimate orientation and heading. They are especially valuable in interior navigation systems where conventional GPS signals are not accessible. The integration of smart sensor technologies facilitates precise localization, navigation, and mapping in diverse fields such as transportation, robotics, augmented reality, logistics, and smart cities. This integration enhances operational safety and efficiency in both indoor and outdoor environments.

6.1.2 Importance of smart sensors in localization, navigation, and mapping

Smart sensors are of paramount importance in the fields of localization, navigation, and mapping owing to their capacity to collect real-time data pertaining to the immediate surroundings and furnish precise positional information (Pahlavan et al., 2015). There are several significant rationales for the importance of smart sensors in various domains.

6.1.2.1 Enhanced precision

Intelligent sensors such as GNSS receivers, IMUs, and LiDAR sensors provide exceptional accuracy in determining location, velocity, and orientation. Accurate localization is of utmost importance in applications such as autonomous cars, since it plays a crucial role in ensuring safe and efficient navigation.

6.1.2.2 Enhancing resilience in complex environments

Smart sensors are specifically engineered to exhibit robustness in demanding situations that may pose difficulties for conventional sensors. IMUs have the capability to offer uninterrupted navigation data in situations when GNSS signals encounter obstructions, such as in tunnels or urban canyons.

6.1.2.3 Processing of data in real time

The integration of computational capabilities in smart sensors enables real-time data processing, resulting in reduced latency and enhanced response times. Time-sensitive applications such as emergency response systems and dynamic routing in transportation heavily rely on the criticality of real-time processing.

6.1.2.4 Incorporation of mapping technologies

The integration of smart sensors with mapping technologies, such as LiDAR-based mapping systems and digital maps, is a common practice. The integration facilitates the concurrent processes of localization and mapping (SLAM), wherein the sensor not only ascertains its own spatial coordinates but also generates a comprehensive representation of the immediate surroundings.

6.1.2.5 The augmentation of safety and efficiency

Smart sensors play a crucial role in enhancing safety and efficiency within the transportation and logistics industry. These sensors offer precise positional information, which is vital in facilitating vehicle navigation, optimizing routes, implementing accident-avoidance systems, and managing traffic effectively.

6.1.2.6 Indoor localization

In situations where GPS signals are either missing or unreliable, indoor localization is facilitated through the utilization of intelligent sensors such as RFID, ultrasonic sensors, and Bluetooth beacons. These sensors provide the implementation of various applications, including asset tracking within warehouses, indoor navigation within malls, and location-based services throughout smart buildings.

6.1.2.7 The incorporation of IoT technology

Smart sensors are essential components of IoT systems as they gather and communicate location data to IoT platforms for the purpose of analysis and decision-making. The integration of location data facilitates the implementation of smart city programs, environmental monitoring, and personalized services.

6.2 FUNDAMENTALS OF LOCALIZATION

In common usage, the terms "localization" and "positioning" mean the same thing: finding out exactly where a gadget is. Tracking, on the other hand, involves the monitoring of moving users or objects, which goes beyond sequential localization and encompasses more sophisticated spatio-temporal processing (Khoshelham and Zlatanova, 2016). The categorization of fundamental localization techniques is determined by the location-dependent measurements that underlie them.

6.2.1 Proximity

The user's location is determined by the transmitter's known location, which is linked to the user's mobile equipment, known as the mobile station (MS). The Cell-ID technique, specified in global system for mobile communication cellular networks, is a representative method that provides the user with the location of the providing base station (BS). A comparable methodology is employed in RFID and Bluetooth-based systems, wherein the assumption is made regarding the proximity of the nearest transmitter (Braun et al., 2015).

6.2.2 The angle of arrival (AOA)

Here, the user's location is approximated by measuring the arrival angles of radio signals broadcast between the MS and various BSs. The method relies on simple geometric connections. Triangulation is the word that describes this approach (Margiani et al., 2023). To determine location in two dimensions, you need at least two BSs; however, you can measure AOA with directional antennas or antenna arrays.

6.2.3 Signal strength

The measurement of strength of signal at the multiple station can be utilized to approximate the distances from neighboring BSs using mathematical models, referred to as path loss models, which depict the reduction in signal strength as a function of distance. The user's position can be determined from the intersection of circles, as each distance represents a circle where the user may live. In order to accurately determine the two-dimensional (2D) user position, it is necessary to obtain Received signal strength (RSS) measures from a minimum of three BSs. This process is commonly referred to as trilateration. However, if additional measurements are accessible, the term multilateration is employed (Lee and Buehrer, et al., 2011). This has the potential to introduce considerable inaccuracies in the calculated distances, leading to inaccurate user location. Fingerprint matching, also called scene analysis, has become widely used as a solution to address the above limitations. This process involves the capture of location-stamped signal signatures, also referred to as fingerprints, at specified locations. Such fingerprints are then stored together with the associated location information in a database, usually in the form of a radio map (Alanezi et al., 2021). Identification of location can be made by using pattern recognition techniques to determine the best possible match between the detected fingerprint at the mass spectrometer and the fingerprints in the radio map. In this case, increased accuracy can be obtained by spending more time and effort in data collection and filling up the radio map to cover the region of interest.

6.2.4 Signal propagation time

The measurement of Time of Arrival (TOA) involves transmitting a signal from the Mobile Station (MS) and receiving it at numerous Base Stations (BSs). This measurement allows for the estimation of distances from the respective BSs by multiplying the distance by the speed of light. Hence, every TOA measurement yields a circular shape, and the aforementioned lateration method can be utilized to ascertain the precise location. In an alternative approach, the measurement of Time Difference of Arrival (TDOA) can be conducted by detecting the received signal at various pairs of Base Stations (BSs). The TDOA measurement delineates a hyperbolic shape, as opposed to a circular shape, through which the user can be situated, with the focal point positioned at one of the two base stations (Patwari et al., 2005).

6.2.5 Smart world and localization

When it comes to the smart world, localization is an integral part of it, and there are many interdependent links between the various parts. All four of these domains—physical, social, cyber, and thinking—make up what is known as the smart world. The subsequent section elucidates the correlation between localization and each of the aforementioned realms (Pahlavan et al., 2015). The physical world encompasses a combination of individuals and objects that are interconnected with these individuals, as well as the surrounding circumstances of their interactions. Undoubtedly, our inquiry pertains to the geographical and temporal positioning of these individuals in relation to the entities they are affiliated with. Given the perpetual movement of both people and things, we are also interested in exploring the connections between place and time in order to examine how one entity may need to relocate in order to have a good interaction with the other. Instances of such interactions encompass intelligent automobiles, potentially autonomous, that necessitate location data in order to traverse roadways (Liang et al., 2016). This may encompass intelligent entities that furnish humans with data based on their whereabouts and circumstances, such as the arrival of buses, taxis, and shared cars. Alternatively, it could entail the utilization of a cartographic representation illustrating different pricing options for an individual seeking information regarding nearby retail establishments offering a specific brand of a desired product. The aforementioned examples serve to demonstrate the clear correlation between individuals and objects. In the social dynamics domain, we see the relationships between objects and individuals, specifically the interaction between individuals and groups. Certainly, the study of social behaviour requires comprehension of the temporal positioning of objects and individuals in relation to one another in order to track their interactions. Representative examples include location-based games like geocaching or finding close acquaintances. Connections among people and objects can reveal powerful people or things, show the diffusion of influence or contagions, and exhibit the stability of these systems. In the virtual world of cyberspace, the application of location and tracking information on individuals and things is used to produce a term referred to as location intelligence (Khelifi et al., 2019). This type of intelligence is developed by observing both the absolute and relative locations of these objects and individuals. An example is the use of web searches on the internet, where it is possible to receive location-based results, commonly referred to as local search results. In thinking, location intelligence is used to improve the quality of life for both human beings and other living organisms. Hence, localization touches upon every aspect of the rising smart world.

6.3 LOCALIZATION TECHNOLOGY AND ITS APPLICATIONS IN THE SMART WORLD

The term "smart world" has gained popularity due to the emergence of various intelligent applications in recent years that utilize time, location, sensor data, and contextual information to address problems. With the introduction of advanced processing capabilities in mobile phones, such as the iPhone, we started referring to them as "smart phones." The advancement of artificial intelligence (AI) in the realm of mechanical devices, facilitated by computer programs and electronic circuits, has led to the designation of these devices as robots (Khelifi et al., 2019). This nomenclature signifies their ability to leverage computational intelligence. In recent times, the term "smart health" has been employed to enhance the analysis of a patient's data, resulting in improved health outcomes and reduced expenses associated with enhancing healthcare services. Buildings that possess a high degree of ubiquitous programmability and intelligence, such as the ability to regulate temperature, are sometimes referred to as smart spaces. It is anticipated that these settings will incorporate several technologies, such RFID tags, Bluetooth, and others, to facilitate the implementation. The term "smart transportation" is employed within the framework of incorporating significant computational intelligence into the domain of traffic monitoring and management. The term "smart grid" refers to the electric system that has the ability to adjust to varying weather conditions and different energy sources, while also intelligently managing electricity use in residential and commercial structures (Shit et al., 2019). Essentially, we are achieving a vision of the "smart world" by developing a range of intelligent applications that possess the ability to address problems. All of the aforementioned apps require location data. This section provides examples of several applications that demonstrate the necessity of localization information.

6.3.1 Localization for smart devices

In order to build smart settings and eventually a smart world, smart devices serve as essential building blocks. Numerous smart devices make extensive use of location data for various purposes. This includes tablets, smartphones, smart watches, smart glasses, and smart televisions. There are two primary categories of these applications. One type is direct applications, which include features like turn-by-turn directions and recommendation systems like Yelp and Kayak for travel prices. Additionally, location information is also indirectly utilized in applications related to gaming or customer behavior analysis (Mittal et al., 2018). The applications necessitate a wide range of accuracy and precision, varying from millimeters in gaming, meters in indoor geolocation, tens of meters in turn-by-turn guidance, and hundreds of meters for broadcasting adverts in specific locations. The smartphone is often regarded as one of the most popular smart gadgets, with a significant majority of the numerous applications specifically built for mobile devices. Smartphones are equipped with various location sensors (Gu and Ren, 2015).

6.3.2 Localization for smart moving platforms

Robotic platforms constitute a crucial element within the dynamic landscape of the intelligent world. A wide variety of fields can benefit from these robots, including medical care, corporate transportation, producing goods, safety, managing warehouses, and photography from the air. Each robot is in motion and requires a positioning and mapping system to determine its whereabouts. Land robots typically utilize a 2D map as their visualization platform, whereas flying robots require a more intricate 3D localization and mapping system (Panigrahi and Bisoy, 2022). Robot localization systems require knowledge of a landmark as the starting or ending point, as well as a technique to monitor the robot's motion. Robotic platforms are equipped with various location sensors such as cameras, speedometers, radio-frequency (RF) communication devices, RFID readers, and optical measuring meters (Lee and Hashimoto, 2003). For instance, a robot functioning within an enclosed laboratory setting or a micro robot in flight may require centimetric or millimetric precision for specific tasks, as opposed to the typically perceived accuracy of a few meters for indoor operations (Čech et al., 2022).

6.3.3 Localization in the context of smart health

Due to the lack of direct visibility into the human body and the presence of several intricate organs, each requiring specialized medical attention, the term "localization" is frequently employed in medicine to identify the precise location of lesions, tumors, bleeding, or pain within the human body. Furthermore, within the hospital setting, there exists a wide array of localization applications that encompass several aspects. These applications encompass the identification of doctors, patients, nurses, and visitors, as well as the localization of routinely utilized equipment like wheelchairs and specialized instruments such as surgical equipment within the confines of the operation room (Coronato and Esposito, 2008). The present surge in research on monitoring the health of the elderly through the utilization of sensor networks necessitates the widespread implementation of geolocation technology for patients. The application of indoor geolocation science and technologies is mostly focused on locating individuals or things within hospitals. Various technologies are frequently employed for indoor and outdoor localizations in the context of health monitoring, particularly where the tracking and location of patients are of utmost importance. According to Paolini et al. (2019), the amount of precision and accuracy needed for localization in the human body can vary from centimeters in the gastrointestinal tract to fractions of a millimeter for neurons in the brain. Applying the same concepts to health-related applications as to intelligent gadgets and robotics ensures accurate localization of people and equipment. In the field of in-body localization and mapping, various imaging techniques have been employed, including traditional 2D X-ray, as well as more modern advancements such as 3D ultra-sound imaging. In recent times, there has been a growing utilization of RF signals and hybrid RF and imaging approaches for the purpose of microbot localization within the human body (Abdellatif et al., 2019).

6.3.4 RFID-based smart spaces and localization

RFID tags are a crucial component of the developing intelligent environments. Passive RFID tags are presently employed in various widely utilized applications, including the tagging of newborn infants or patients in medical facilities, the verification of vehicles at highway toll stations. The widespread deployment of RFID tags in smart buildings and the integration of RFID readers in future smartphones are anticipated. In the future, we anticipate the utilization of trillions of RFID tags for a wide range of applications in the intelligent world (Moreno et al., 2012). The necessity of localizing RFID tags is increasingly crucial, alongside the significance of connecting these tags via the IoT. Passive RFID readers are uncomplicated proximity tracking technologies that transmit tag information to the cyber domain (Buffi et al., 2021).

6.3.5 Localization in the aspect of smart transportation systems

Smart transportation plays a crucial role in the development of smart cities and the broader context of the smart world. Currently, there is a comprehensive understanding of the precise geographical coordinates of ground and air transportation systems in outdoor settings. If effectively interconnected, this data has the potential to significantly enhance the efficiency of transportation networks, leading to substantial reductions in both fuel and transportation expenses. Vehicular navigation systems predominantly depend on GPS and mechanical sensors to ascertain the velocity and trajectory of the vehicle (Zhao, 2000). One of the limitations of existing vehicular systems is the absence of a robust networking infrastructure that enables efficient coordination of these movements. This is a lengthy procedure that entails the establishment of networking standards. Another growing application of vehicle localization involves determining the relative location of cars in relation to nearby mobile and stationary objects. It is imperative to develop an application that can aid drivers in effectively managing vehicle movements in relation to both other moving vehicles and the stationary infrastructure in their vicinity. Given the substantial investments made in the numerous sensors deployed on mobile vehicles and their associated infrastructures, it is desirable to possess knowledge regarding their precise locations and the corresponding traces. This information encompasses the mobility patterns of individuals who are constantly changing, and it facilitates the development of various applications based on data mining.

6.3.6 Localization for smart infrastructure

The power grid, intelligent transportation networks, and water systems are just a few of the critical national infrastructures that have been increasingly incorporating smart technologies in recent years. For critical infrastructure resource management, resilience, and seamless operation, localization is essential. Even though there is not a ton of studies on topics like smart grid fault localization, it is vital to find out exactly where resources are being cut or added. Because it has a direct effect on the measured quantities and the nearby infrastructure, sensor placement is critical for water quality sampling and electric grid load evaluation (Winter et al., 2019). There are a number of prerequisites for thinking about precision and accuracy. Knowing the locations of various components down to a degree of accuracy of tens of meters on a map may be adequate at a macroscopic level. The localization precision may, however, need to be on the centimeter scale within a particular component, such an electrical substation.

6.4 NAVIGATION TECHNIQUES

6.4.1 GNSS-based navigation

The global navigation satellite system (GNSS) is a system of interconnected satellites that provide precise location and navigational data to users all over the world. For precise location and timekeeping, the world relies on GNSS, a global system. In the social dynamics domain, we see the relationships between objects and individuals, specifically the interaction between individuals and groups. Certainly, the study of social behavior requires comprehension of the temporal positioning of objects and individuals in relation to one another in order to track their interactions. Representative examples include location-based games like geocaching or finding close acquaintances. Connections among people and objects can reveal powerful people or things, show the diffusion of influence or contagions, and exhibit the stability of these systems. In the virtual world of cyberspace, the application of location and tracking information on individuals and things is used to produce a term referred to as location intelligence (Khelifi et al., 2019). This type of intelligence is developed by observing both the absolute and relative locations of these objects and individuals. An example is the use of web searches on the internet, where it is possible to receive location-based results, commonly referred to as local search results. In thinking, location intelligence is used to improve the quality of life for both human beings and other living organisms. Hence, localization touches upon every aspect of the rising smart world.

6.4.2 Global positioning system (GPS)

GPS is a widely recognized GNSS that has been designed and is being employed by the government of the United States. A network of satellites called GPS orbits the planet, sending out signals that can be picked up by receivers either on the ground, in the air, or in deep space. The global location system (GPS) offers precise location, navigation, and timing data to users around the globe. Below are several crucial aspects about GPS.

6.4.2.1 Segments

The GPS system is partitioned into three distinct components. The space segment encompasses the satellites that are currently in orbit. The Control Segment comprises terrestrial monitoring and control stations that oversee and regulate the satellites, guaranteeing their precision and dependability. The user segment includes GPS receivers and user devices that utilize satellite signals for the purpose of determining locations, velocities, and timings.

6.4.2.2 Constellation

At least twenty-four satellites make up the GPS, and they are all situated in medium Earth orbit (MEO), some twelve thousand miles (12,550 km) from the surface of the Earth. Any point on Earth may see at least four of these space craft at any one time thanks to the way they are structured.

6.4.2.3 Signals

GPS satellites consistently emit signals that encompass data regarding their geographical coordinates and the exact moment at which these signals were transmitted. The aforementioned signals propagate at the velocity of light and are detected by GPS receivers situated on the Earth's surface or within diverse equipment.

6.4.2.4 Trilateration

It is a technique employed by GPS receivers to ascertain their precise location. The technique of trilateration entails the measurement of the temporal duration required for signals originating from numerous satellites to reach the receiver. The receiver may determine its own position by utilizing the distance to each satellite, which is determined by scheming the signal travel time and the speed of light.

6.4.3 GLONASS

The global navigation satellite system (GLONASS) is a satellite navigation system (GNSS) that has been created by Russia, bearing resemblance to the United States' GPS system. GLONASS comprises a network of satellites that revolve around the Earth. The primary objective of the GLONASS constellation is to establish a total of 24 operational satellites, supplemented by a number of spare satellites. Similar to GPS, GLONASS offers geolocation, navigation, and timing services on a worldwide scale. The technology was primarily designed for military and civilian applications, encompassing a wide range of uses such as vehicle, ship, and airplane navigation, as well as precise time for scientific and commercial endeavors. GLONASS operates based on principles that are analogous to GPS (Revnivykh et al., 2017). GLONASS receivers on the ground or in various devices receive signals continuously transmitted by the satellites in the constellation. These signals encompass data regarding the positions of satellites and the precise moment at which the signals were broadcast. Trilateration is employed by GLONASS receivers in order to ascertain their precise location. The receiver can determine its position by monitoring the latency of signals from numerous GLONASS satellites and utilizing the known positions of the satellites.

6.4.4 Galileo

Galileo is GNSS operated by the European Union. The primary objective is to augment Europe's self-sufficiency in satellite navigation and diminish dependence on non-European systems such as GPS. The Galileo constellation consists of many satellites suspended in MEO at an approximate altitude of 23,222 km (14,429 miles) above the Earth's surface. According to the most recent update, the Galileo system intends to possess a cumulative count of 30 functional satellites, encompassing additional spares. Galileo operates through the utilization of satellites that emit signals, which are subsequently detected by Galileo-compatible receivers situated on the Earth's surface or within diverse equipment (Falcone et al., 2017). These signals encompass data regarding the placements of satellites and accurate timing information. The position of Galileo receivers is determined through the application of trilateration methods. The receiver precisely determines its own position by detecting the latency of signals from numerous Galileo satellites and utilizing the satellites' positions.

6.4.5 BeiDuo

BeiDuo is a prominent GNSS owned by China. It offers positioning, navigation, and timing services that are comparable to those provided by other prominent GNSS systems. The BeiDou system was created with the purpose of offering China and the Asia-Pacific area a self-governing GNSS that provides worldwide coverage and precise navigation and timing services

(Rajavarathan et al., 2024). The objective is to augment China's strategic autonomy in satellite navigation and diminish dependence on foreign global navigation satellite systems.

6.4.6 NavIC

Indian space research organization (ISRO) created NavIC with the purpose of delivering precise location, navigation, and timing services across India and its neighboring area. The primary objective of this initiative is to provide autonomous satellite navigation skills to many industries, encompassing transportation, agriculture, crisis management, and defense. The NavIC system consists of a collection of seven satellites that are positioned in geostationary and geosynchronous orbits along the Earth. The satellites have been strategically deployed to provide coverage for the Indian subcontinent as well as an expanded region spanning up to 1,500 km beyond the borders of India.

6.4.7 Dead reckoning navigation

Several data, including position, speed, acceleration, and orientation, can be measured using sensors in this process. The bulk of self-driving vehicles move around their working environment using wheels. The simplest way to pinpoint their exact position is to monitor the wheel spin speed and steering system alignment. The next step is to use this information with the vehicle's kinematic model to determine its predicted location.

To find out how fast a wheel is turning, a rotating sensor is usually fastened to the axle or the motor shaft. The resolution, maximum speed, and cost considerations will determine the final sensor choice (Jirawimut et al., 2003). When it comes to dead reckoning, a lot of common tools are potentiometers, incremental and absolute encoders (e.g., optical, inductive, capacitive, magnetic), tachometers, synchros, resolvers, compass, and inertial sensors like accelerometers and gyroscopes. With an emphasis on inertial sensors, this section offers a thorough review of the most common dead reckoning sensors.

6.4.8 Potentiometers

The sensor in question operates on a variable resistor that exhibits a value that varies in accordance with the orientation of its shaft. There exists a wide range of potentiometer types, encompassing both single and multiple turns. The measurement of the analog voltage between the moving contact and one of the extremes of the potentiometer is used to sense the position. The resolution of the sensors is contingent upon several factors, including the device's quality, the quantity of electronic noise, and the number of digits employed by the analog to digital converter (Halvorsen, 2021).

6.4.9 Encoders

The aforementioned sensors are widely employed for the purpose of quantifying the velocity or location of a spinning apparatus. In order to quantify the rotational velocity, encoders are commonly affixed to a shaft. The sensors exhibit band-pass frequency characteristics. For dead reckoning to work, the forward kinematic equations of the vehicle are sent into the wheel rate encoders and the steer angle encoders, allowing for the prediction of the vehicle's direction and position. One drawback of encoder-based dead reckoning is that it often uses a 2D vehicle model, which leads to less precise results when the terrain moves away from a plane (Cho et al., 2010). Another significant issue pertains to the inability of encoder-based dead reckoning to accurately identify wheel slip.

6.4.10 Gradual-based encoders

Sensors of this nature are founded upon a diverse range of principles. Typically, these devices possess one or two channels that are utilized for the purpose of measuring or quantifying the diameter of a shaft. The device's sensitivity to minute shaft rotations increases proportionally with its resolution. These devices usually produce square-wave signals that are proportional to the number of markings found on the shaft. One-channel encoders can only detect direction and deliver velocity data. An extra square wave with the right phase shift, generated via an extra channel, fixes this problem. While the sensors work well at medium and high frequencies, they might produce significant errors at very low ones. As a rule, the system depends on the sensor and a decoder to transform the numerical representation of the square-wave output. The most common and extensively used encoders are based on optical principles. In order to block the light from two optical devices, the sensors have a see-through disk decorated with many lines. The devices are attached in a way that produces a quadrature output, which is a square wave with a phase shift of 90°. On top of that, when a revolution starts, an index pulse is generated. In respect to a starting point, the aforementioned sensors can measure the relative size of a shaft. Increasing the number of lines on the see-through disk could improve the resolution. As a function of disk diameter and light transmission quality, the maximum number of counts each rotation is determined by the encoder. A maximum of 10,000 counts per rotation can be achieved with an encoder with a diameter less than 3 inches.

6.4.11 Encoders with absolute values

These gadgets provide precise data regarding the precise location of the shaft. These devices have the capability to be built with resolutions ranging from 8 to 10 bits per revolution. These devices find utility in scenarios where achieving high levels of accuracy is not critical and when the loss of reference due to power outages is deemed unacceptable. High-resolution devices can be produced with identical environmental specifications as incremental encoders, albeit at a significantly higher cost compared to their counterparts.

6.4.12 Orientation

These instruments are utilized for the purpose of quantifying the orientation of the Earth's magnetic field. The utilization of this physical theory for navigating ships across the ocean dates back to the 1200s. An uncomplicated apparatus comprised a metallic needle suspended in water. The accuracy of the compass data necessitates rectification due to the disparity observed between the true north and the orientation of the Earth's magnetic field. An additional adjustment is necessary as a result of vertical magnetic forces. The magnitude of this correction is contingent upon the geographical location of the vehicle, with minimal impact at the equator and maximal impact at the poles.

6.5 IMPULSE SENSORS

The interior status of the vehicle is measured via inertial sensors. An important benefit of inertial sensors is their lack of radiation and susceptibility to interference, as well as their ability to be enclosed and protected from the surrounding environment. This characteristic renders them potentially resilient in challenging environmental circumstances. Accelerometers and gyroscopes are widely recognized as the prevailing types of inertial sensors. The measurement of acceleration in relation to an inertial reference frame is conducted by accelerometers. Both gravitational and rotational acceleration, along with linear acceleration, are encompassed within this category. Gyroscopes quantify the rotational speed without considering the coordinate system. In addition, they can offer 3D position data and, unlike encoders, have the capability to detect wheel slip (Li et al., 2021). An further utilization of inertial sensors involves the utilization of accelerometers for the purpose of quantifying the vehicle's attitude. Two orthogonal accelerometers can be used to assess the tilt of a platform. A comprehensive understanding of the gravity value inside the operational zone is necessary for this task.

6.5.1 Millimeter-wave radar

Radar allows for the measurement of the distance, direction, and velocity of objects. It is a significant component inside contemporary advanced driver assistance systems, with the purpose of detecting blind spots, facilitating lane changes, and issuing rear/forward cross-traffic alerts. An affordable micro-radar has the capability to produce several "pings" that enhance both the range and sensitivity. Radar has the capability to identify objects located in close proximity to a vehicle, including those in front, to the side, and to the rear, in the majority of weather circumstances (Jha et al., 2019). Nevertheless, the industry is currently facing issues regarding the potential for false positives in the received image due to the presence of numerous moving vehicles operating on the same radar frequencies in close proximity. Radar serves as a valuable addition to GNSS-based Position navigation and timing (PNT) in the near field, but it cannot completely replace it. These gadgets have the capability to ascertain the distance and direction to a collection of beacons positioned at predetermined sites. These targets facilitate the direct reflection of the incident radiation beam along its original path, spanning the capture angle of the beacon. The aforementioned sensor has been employed in the autonomous navigation of a truck with a gross weight of 80 tons, achieving precision levels of centimeters at speeds of up to 6 ms/h. Despite its higher cost compared to a similar GPS system, this sensor offers enhanced reliability due to its ability to provide complete control over both the quantity and positioning of beacons inside the operational region. Contrary to laser units, which usually provide information on the distance and direction to the nearest reflector, radar systems provide power spectra for each azimuth (Wang, 2016). The aforementioned characteristics render this sensor highly appealing for dependable outdoor navigation.

6.5.2 LiDAR-based navigation

LiDAR is a technique used to measure distances. Essentially, it is a variant of radar that employs light instead of radio waves. A LiDAR device comprises a range measurement sensor that generates a light pulse in a repetitive manner. Once the light interacts with a designated point, it undergoes reflection and then reaches the sensor. The sensor then ascertains the distance to the object by measuring the duration it takes for the light to traverse the target and subsequently return. Furthermore, the placement of the range measuring sensor on a rotating platform facilitates the device's ability to capture measurements at various locations spanning a whole 360° (Malavazi et al., 2018). The sensor rapidly captures range measurements (up to around 10,000 samples per second) as it rotates, offering a comprehensive 2D perspective of the robot's environment. A raw map is obtained by conducting a 360° view sweep and collecting several range samples. The subsequent stage in this procedure involves integrating the 360° scans and consolidating them into a comprehensive map. When the robot navigates within its surroundings, it possesses the capability to ascertain its position in relation to the current and previously acquired data through the process of localization. Subsequently, it conducts more scans and incorporates them into the existing map. The term "Simultaneous Localization and Mapping"

(SLAM) originated from this particular procedure. Users are presented with a sequence of distance measurements at various angles and/or directions, which are facilitated by systems that manipulate the laser in a plane or another pattern. By utilizing the provided 2D or 3D range data, individuals have the capability to transform radial coordinates into Cartesian coordinates, so enabling the generation of a 2D or 3D contour map of the scene.

6.6 NAVIGATION TECHNOLOGIES FOR VISUAL IMPAIRMENTS

Dot Walker: It is an Android application designed for outdoor navigation, specifically catering to the requirements of travels without sight. The utilization of the TalkBack screen reader is anticipated in conjunction with this application. Dot Walker employs several features such as direct speech output, audible alerts, vibrancy, and high-contrast writing that may be adjusted to grow as required. Dot Walker Pro is the name of the premium version (Messaoudi et al., 2022). The Nearby Explorer program is a GPS-based tool that has been purposefully developed to cater to the requirements of those who are blind or visually impaired. This cutting-edge software not only offers guidance, but also utilizes several methods, including sign interpretation and route parameters, to provide a comprehensive and intricate depiction of the surroundings. By utilizing internal maps, there is no necessity to establish connections between information. However, when a map is available, Nearby Explorer enhances the internal map data by incorporating places obtained from Google Places. The free edition of Nearby Explorer Online d offers a reduced range of functions and does not incorporate internal maps. The application will enhance the information by including Google location data when mobile data is connected (Isazade, 2023). OpenStreetMap allows users to build, name, and share favorites, enabling others to utilize the locations they have marked and identified. The application may utilize its location and orientation to acquire objectives, such as the trajectory of a vehicle, and can also function as a compass.

6.6.1 GetThere

It is a specialized assistive technology created for those with visual impairments, rather than being a modification of technology intended for sighted individuals. The application operates in conjunction with TalkBack, the widely used Android screen reader, in a manner that ensures the absence of conflicting signals. The GetThere application does not display a map, but rather provides information about local location and navigation instructions. The navigational guiding system provides audio cues both prior to and following each intersection. At any given moment, you have the option to request GetThere to provide your current location by simply shaking your mobile device. GetThere is a tool that identifies deviations from the intended course and provides information on the specific errors made, as well as the necessary steps to return to the intended path (https://docs.google.com/document/d/193E0d2bN6).

6.6.2 Google Maps

Google Maps introduced a new tool called detailed voice guidance in October 2019. This program provides audio directions to aid blind and visually impaired folks while walking. This feature signifies a notable progression in the accessibility of Google Maps, specifically for persons who have visual impairments. In order to avail themselves of this functionality, users are need to open their Google Maps settings and select the "Navigation" option. Following this, it is necessary to choose the option labeled "Voice guidance" from the Options category situated in the lowermost section of the list (Sugiyama, 2020).

6.6.3 Square of blindness

The iOS smartphone is equipped with a GPS and compass functionality, enabling users to determine their location and collect data about the surrounding online environment.

6.6.4 Guide for Low Viz

This program utilizes location services to facilitate users in navigating extensive regions and ascertaining the optimal route between two given sites. The Low Viz guide incorporates the following elements, as navigation will be accessible via the smartphone capability. The topics of interest include indoor location-based and presence recognition, navigation and pathing, and locations sharing. A link that is based on location available to track the location.

6.6.5 OSM for individuals with visual impairments

OpenStreetMap (OSM) has undertaken initiatives with the objective of providing OSM services that are specifically designed to cater to individuals who are blind or visually impaired. According to Wiki (2020), these projects entail teams of specialists who employ various methods to create web-based instructions that are accessible to individuals who are blind or visually impaired, using maps. An example of such an endeavor is the creation of the online platform https://www.openstreetmap.org/traces/tag/blindOSM, which provides tactile maps that may be generated or printed on microcapsule paper or 3D printers. Users can utilize shortcut options to expand the maps, which showcase common locations like buildings and entrances.

The inclusion of voluntary geographical information is a crucial element within the OSM framework, since it facilitates the development of customized solutions for individuals who are blind or visually impaired. Furthermore, individuals who suffer color blindness have reported a higher level of usefulness when using OSM in comparison to Google Maps. Individuals who possess a color combination of red and green exhibit enhanced performance compared to those who possess the color red. Conversely, a considerable portion of the population finds OSM to be more challenging to navigate in comparison to Google Maps.

6.6.6 BrailleNote GPS

The GPS receiver is a crucial tool for position-based mapping, as it provides essential location information required for autonomous navigation among those who are visually impaired or have partial vision. This program enhances the process of trip preparation by providing users with the ability to choose any destination throughout the globe and journey there it either in person or through virtual means. In addition, the GPS device has the capability to detect and automatically announce nearby destinations.

6.6.7 Blavigator

It is a navigation system. Blavigator, also known as Blind navigator, is the next iteration of the smart vision project and aims to further enhance and optimize its functionality. The present system incorporates a collection of specialized modules, with each module assigned distinct functions. There are three main components that make up the interface: text-to-speech technology, a command pad with four buttons, and vibratory outputs. Text-to-speech technology has a twofold function: it effectively communicates essential information to the user and facilitates the navigating of menus (Adao et al., 2012). The system can be accessed by the user through the manipulation of a command pad consisting of four buttons. When the system requires heading instructions such as "turn to the left" or "turn to the right," the vibratory devices surrounding his body will be activated. Specifically, the researchers explored the ability of this map to facilitate independent and efficient movement within a given region, as well as the ability to perceive and locate specific points. The determination of whether a specific area is depicted on the intended map or not.

6.6.8 Virtual reality

In a study conducted by Picinali et al. (2014) the focus was on assisting individuals with visual impairments in navigating unfamiliar environments through the implementation of training techniques that enhance auditory perception and interaction with virtual reality events. In this study, a comparison was made between two learning methods: in situ mobility and navigation in virtual architecture using audio input exclusively. However, notable distinctions were noted between the two environments that were examined, namely in terms of the topological and metric characteristics of mental pictures acquired during real-world navigation.

6.6.9 Guide for blind individuals

An innovative methodology for identifying and alerting persons of obstacles by employing an ultrasound-based regional network. This strategy has promise as a viable and all-encompassing solution for those with visual impairments as they navigate various environments. Based on a comprehensive analysis of costs and time, it has been determined that this particular technology is a suitable and feasible solution for the visually impaired population (Červenka et al., 2016).

6.7 SMART AGRICULTURE: MANAGING THE DIGITAL FRONTIER

Agriculture is fundamental to human existence and the development of our civilization since it produces food, fuel, fiber, and other necessities. Nevertheless, the agricultural sector is facing significant threats from human-caused climate change, which threatens both food and nutritional security. Global temperature fluctuations, environmental degradation, a lack of agricultural labor, an aging population, and changes in food consumption habits are just a few of the farming difficulties that the agricultural sector must adapt to. A wide range of approaches, including precision agriculture (PA), genetic engineering, and plant breeding, have been used to increase agricultural output. According to Lima et al. (2021), the use of drones, smart sensors, robots, and remote sensing has enabled a surge in new scientific discoveries and technical advancements globally in the past few years. Because of this, a plethora of new methods have emerged for various purposes; these include high-throughput phenotyping, agricultural practices, disease detection, forecasting, yield prediction, and weather prediction. These environmental practices are crucial to crop growth, yield, and quality. In addition, there are a number of elements that have contributed to the present wave of digitalization, including the development of analytics, the rise of big data, the fall in the cost of technology, better internet access, and increased computational capabilities. The potential for commercial agriculture to achieve smart farming, resilience, productivity, and sustainability is enhanced by these characteristics taken together. Technological advancements allow for the efficient monitoring of soil, crops, and environmental factors over large areas. This data is then used by farmers to improve yield while reducing environmental impacts through careful management practices (Ehlers et al., 2021). The huge costs of implementation, worries about data security, and a lack of digital literacy among farmers are just a few of the challenges that smart farming faces, despite its great potential.

6.7.1 Monitoring and phenotyping of crops

AI in agriculture encompasses the utilization of AI to augment agricultural productivity and facilitate real-time monitoring, harvesting, processing, and sales activities. It is playing a pivotal role in transforming the agricultural sector. AI plays a crucial role in safeguarding the agriculture sector from a range of challenges, encompassing population growth, climate change, employment scarcity, and food security (Saiz-Rubio and Rovira-Más, 2020) has played a significant role in enhancing crop output and facilitating real-time monitoring, harvesting, processing, and selling processes.

6.7.2 Enhancing crop production and breeding with precision agriculture and remote sensing technologies

Precision farming is widely recognized as a crucial and sustainable methodology for optimizing crop yield and enhancing the socioeconomic and livelihood circumstances of farmers. PA encompasses a range of methodologies that find applications across several agricultural research domains. This is an innovative approach to managing agricultural systems that utilizes georeferenced data. It involves monitoring procedures and combines soil, plant, and climate factors to provide detailed georeferenced data. The goal is to provide information that can inform more informed decision-making. Moreover, following a thorough analysis of the various factors that impact crop growth within a specific agricultural area, it advocates for the implementation of tillage practices and the precise utilization of agricultural inputs, including herbicides, fertilizers, and irrigation. PA employs sophisticated technology, sensors, and analysis tools to facilitate management, decision-making, and enhancements in crop output. The global community has enthusiastically adopted PA, an innovative concept that holds the potential to enhance productivity, reduce personnel expenses, and ensure effective administration of irrigation and fertilizer systems. Improving crop quality, raising yields, and making the most of agricultural resources are all goals that see heavy data and information utilization. This cutting-edge technique has recently gained popularity among farmers looking to maximize their return on investment from resources like water and fertilizer by increasing output while simultaneously improving product quality. It is critical to collect a large amount of high-resolution data about the crop's state or health during the growing season. By combining GPS with geographic information system (GIS), a vast amount of geospatial data can be more effectively handled and analyzed, which has led to advancements in site-specific agriculture. Using GIS and GPS improves PA by providing accurate location data and the ability to conduct spatial analyses. Accurate navigation is made possible by GPS technology, which improves field operations, and variable rate technology, which adapts input application to field fluctuations, is made possible by the GIS system. Applications based on GPS are already being used in many areas of agriculture, including field planning, soil assessment, crop exploration, variable rate application, agriculture, and yield mapping. Furthermore, most PA application technologies rely on GPS systems, as clients use them to remotely coordinate manufacturing operations at specific locations in real time (Mogili and Deepak, 2018). They find their primary application in the agricultural sector, where they serve many functions such as soil monitoring and mapping, field contouring, and production tracking. In a typical field-traveling vehicle, one can find a GPS receiver, an information retrieval unit, mapping and visualization software, and a differential GPS. The farmer utilizes the same equipment to delineate the boundaries of the field and gathers data using both GPS units. In addition, they have the capability to navigate the region and document information regarding areas impacted by pests and diseases throughout the growing season. During the process of crop surveying, GPS systems are utilized to navigate farm trucks within certain field regions. The rates at which a farm vehicle may perform chores vary depending on the soil conditions and production system parameters in each zone.

6.7.3 Agricultural automation and robotics

Automation has had a profound impact on the agricultural sector, since it has facilitated the ability of machines to carry out tasks that were previously reliant on human labor. The integration of technology in agriculture is important in response to the increasing global population and growing food demands. One advantage of autonomous farming is the utilization of intelligent seeding robots capable of identifying crop varieties and planting zones. This technology enhances operational efficiency by streamlining the process. Weeding robots facilitate accurate application of herbicides, hence reducing pesticide usage and mitigating environmental consequences (Veroustraete, 2015). Moreover, by employing robotic arms, automated harvesting robots can ascertain the optimal moment for harvesting a crop, thereby diminishing labor costs and time allocated to the activity, while simultaneously enhancing productivity. Robotics and autonomous systems are being deployed in key economic sectors such as agri-food, which exhibit relatively low productivity levels. Robotics has been shown to have a substantial influence on the management and productivity of agriculture. In light of the suboptimal performance exhibited by conventional agricultural gear, scholars have increasingly directed their attention toward the development of autonomous agricultural equipment. The objective of this technology's development is to replace human labor and produce efficient results for facilities of varying sizes. The implementation of robotic technologies inside this particular industry has significantly enhanced output levels. In addition, robotics technologies enhance agricultural practices through the automation of harvesting processes, which involves the identification and analysis of various factors and characteristics of vegetation. These technologies also facilitate monitoring and data collection by providing insights into soil health and environmental conditions (Ahirwar et al., 2019). Furthermore, they aid in sorting and packaging agricultural products, offer labor assistance through the use of exoskeletons and wearables, and conduct robotics scouting activities to detect diseases and nutrient deficiencies in crops. Occasionally, a seemingly inconsequential yet consequential alteration toward the progress of technology yields numerous advantages. An instance of this is the creation of Eli Whitney's cotton gin, which sparked the concept of creating a technology capable of significantly accelerating the separation of cotton fiber from seed, thereby transforming the cotton business. The production of fifty pounds of cotton during a 24-hour period has led to the development of autonomous agricultural robots. An uncomplicated automatic technique was created to precisely ascertain the location of seeds. Furthermore, a highly precise method for seed placement was developed in order to ensure a seed ground velocity of zero. This is of utmost importance as it guarantees that the seed will not undergo rebound upon contact with the soil. The plant's development or status was documented by automated machinery. A multitude of biosensors have been created for the purpose of monitoring plant growth and detecting plant diseases. Laser weeding technology has replaced the human weeding approach by employing a computer-controlled mobile concentrated infrared light beam to harm weed cells (Sparrow and Howard, 2021).

6.8 MAPPING SYSTEMS

In recent decades, there has been a meteoric rise in the demand for precise and continuously updated geographic data. Indoor and outdoor 3D modeling, GIS data creation, disaster-response HD map building, and autonomous car research and development are just a few of the many applications that rely on geospatial data. Constant improvements to mobile mapping systems (MMSs) have made it easier to get this data. According to Otero et al. (2020), a mobile mapping system is a setup that uses a combination of mapping sensors attached to a moving platform to gather geographic data while also determining the platform's position. The main sensors used by a typical MMS platform to collect data about objects or areas of interest are LiDAR and/or high-resolution cameras. The IMU and global navigation satellite system (GNSS) are two examples of the positioning and georeferencing systems that these sensors are a part of. Newer MMSs have aimed to eliminate human error in data collection and processing by utilizing a multi-sensor platform and accomplishing direct georeferencing. Machine learning and AI have been used into automation systems to better extract and map offline and online objects like road signs and traffic signals. Significant algorithmic improvements in photogrammetry, computer vision, and robotics have contributed to the field of mobile mapping technology's meteoric rise in the last several decades. Utilizing location sensors allows for the acquisition of geographical coordinates and movement data, which can then be used to georeference the 3D data that have been collected. Along with GNSS, IMU, and digital measurement instrument (DMI) (more especially odometers), these sensors include a plethora of others. It is standard practice to use fusion techniques to combine measurements from multiple sensors to improve the statistical accuracy of location. In addition, the location and navigation cameras can be fused to perform additional tasks. Since the GNSS receiver and the IMU/DMI alone cannot provide accurate and trustworthy measurements for mobile platform navigation, sensor fusion solutions are now seen as the standard practice for location (Lovas et al., 2020). In different environments, the signal intensity of GNSS measurements can fluctuate. For example, while you might get a strong signal in open spaces, you might get a weak signal or signal degradation in confined spaces, leading to data loss. In contrast, when combined with GPS data, the IMU and DMI can introduce a large amount of errors and are thus often used as supplemental observations for navigation. Data collection sensors rely on LiDAR and digital cameras to provide raw, unprocessed 3D and 2D coordinates of the environment. While LiDAR sensors are used to measure MMSs in three dimensions, photographs are mostly used to provide spectral and colorimetric data. By employing advanced dense image matching algorithms, stereoscopic images have been made possible, paving the way for the collecting of further dense measurements for 3D data fusion. Below, we have laid out a thorough analysis of location and data gathering sensors, along with the related approaches to sensor fusion.

6.8.1 Systems and platforms for mapping mobile devices

Which sensor and platform are best suited to a particular job depends on a number of things. This study takes into account the technical solutions, project budget, and sensor scope. Processing methods and scene attributes (such as indoor/outdoor status, for example). These help identify the accessible platforms (e.g., transportation-mounted or backpack-style) and the available sensors (e.g., those with or without GPS). Interior areas do not have access. Because of this, we need to find ways to replace GPS signals or cars. We have considered, generally speaking, the two primary ways in which MMS platforms are classified: conventional vehicles and the atypical lightweight/portable mapping devices.

6.8.2 Vehicle-mounted systems

Perform most of your work on key routes, collecting 3D data on a city or block scale. Depending on their intended use, nontraditional portable devices like backpack/wearable systems, handheld systems, or trolley-based systems can function indoors or outside in locations without GPS connectivity. In outdoor environments, these mapping approaches are commonly used to augment vehicle-based systems. They are particularly useful for mapping areas that larger vehicles cannot reach, such as narrow streets (Shen, 2013). One component of vehicle-mounted systems is the attachment of sensor suites to vehicles for the purpose of collecting dense point clouds. These devices allow for the quick gathering of data when the vehicle is in motion (20–70 mph). Depending on the mapping application, the sensor platform can be attached to vehicles, trains, or boats. Because of their small size and luggage capacity, vehicle-mounted systems are able to incorporate high-quality sensors, resulting in generally greater accuracy compared to other mobile mapping platforms. The incorporation of survey-grade LiDAR technology into a vehicle-mounted MMS system allows for the provision of accurate and comprehensive measurements. Furthermore, it includes a 360° field-of-view camera that can capture texture data (Fassi and Perfetti, 2019). To find its location, a system mounted on a vehicle usually uses GNSS receiver readings in conjunction with IMU and DMI. Autonomous driving and railway monitoring applications can benefit greatly from these systems, which can also automatically identify changes within the mapped zones, provide accurate maps, and more.

6.8.3 Handheld and wearable systems

They use sensors of modest size and are characterized by being lightweight and compact. The platform can be worn or held by the operator as they move through the specified area of interest. One common feature of wearable technology is the backpack system, which allows users to collect data even when they are on the go. The mobility of both wearable and handheld devices makes them ideal for mapping areas not accessible by traditional GPS methods, such as densely populated areas, areas with complicated topographies, or small, tight locations that are too narrow for cars to navigate. According to Lauterbach et al. (2015), GNSS receivers might not be the best choice for positioning portable and wearable systems in these specific situations. Alternatively, they can gather data and pinpoint their location using technologies like LiDAR and cameras, or an IMU. Some examples of such devices include the HERON LITE Color and the Geo SLAM. In addition to Zeb Revo Go and Zeb RT, other options include Leica BLK2GO and Leica Pegasus:Backpack as well as HERON MS Twin, Nav Vis VLX, and Viametris BMS3D-HD. Previous research has shown that portable and wearable technologies can effectively map indoor areas. One use for these devices is to create maps of caverns that do not have any illumination or GNSS signals. Also, they are used for mapping cultural heritage locations, which can be rather complicated and require a lot of data collected efficiently from different angles. Forest surveys, security and safety maps, and building information modeling are just a few examples of the kind of tasks that these systems excel at efficiently mapping.

6.8.4 Mapping technologies applications

Mobile mapping offers excellent resources for various applications, facilitated by the widespread availability of user-friendly and portable MMS platforms, as well as their adaptability to varied operating conditions. This is especially advantageous because the majority of these applications depend on routinely collected data for the purpose of detecting and monitoring, such as powerline detection and monitoring in train systems.

6.8.5 Road asset management and condition assessment

MMSs that are deployed on roads have the capability to consistently gather precise 3D data of the road and its surroundings. This data greatly aids in the management, mapping, and monitoring of road assets, such as road signs, traffic signals, and pavement dimensions. The creation of road asset inventories is crucial due to the significant quantity of road assets. Moreover, given the growing degradation of road conditions, there is a pressing need for automated methods of routine transportation maintenance, such as the detection of pavement cracks and distress (Stricker et al., 2019). Hence, a primary advantage of creating a current and precisely georeferenced road asset inventory is to enable automated and effective identification of changes, replacing the traditionally time-consuming process of manual inspections. The road condition monitoring processes generally encompass four distinct stages: (1) the acquisition of data through the utilization of mobile monitoring systems, (2) the identification of defects, which can be automated through the application of deep learning techniques, and (3) the evaluation of defects.

6.8.6 Emergency and disaster response

The geospatial data offered by the MMSs play a crucial role in enhancing emergency, disaster response, and post-disaster recovery initiatives. MMSs offer cost-effective and time-efficient methods for gathering and generating 3D reconstructed models that contain comprehensive information on the meaning and structure, enabling navigation during emergencies or disasters. It has been documented that MMSs have the capability to offer building-level data, such as information on floors, walls, and doors, with a resolution and accuracy of centimeters. Oftentimes, the structural blueprints of a building are not current after it is built, which can impede a rescue operation during a fire emergency. However, the MMS can offer a cost-effective solution to generate a precise and current 3D model of a facility or building, thereby facilitating emergency responses. Another instance involves gathering 3D data about roadside assets and inputting it into GIS. These systems can function as instruments for pre-event analysis, enabling the identification of probable consequences of natural disasters through simulations such as flood simulations and earthquakes (Laituri and Kodrich, 2008). This, in turn, assists in the development of preventive planning strategies. The future prospects of MMSs entail the development of more streamlined data collection techniques, such as a person using a cellphone to capture images of their surroundings. While these methods may have reduced accuracy, they have the potential to provide crucial georeferenced information during catastrophic or emergency situations, enabling situational awareness and facilitating the implementation of appropriate remedies.

6.8.7 Digital heritage conservation

The significance of digitally documenting archaeological sites and safeguarding cultural assets is increasingly recognized by individuals. A significant number of these monuments face the risk of degradation and structural failure, a process that can be expedited by adverse weather conditions and natural calamities. For instance, the collapse of numerous cultural structures in Nepal and Iran can be attributed to seismic events. Hence, it is imperative to actively record these locations while they remain intact. Additionally, a thoroughly documented heritage site can facilitate alternative forms of tourism, such as virtual tours, to alleviate the burden of on-site visits and mitigate the human factors that contribute to the degradation of these sites (Malinverni et al., 2018). MMS has been utilized as a key source for acquiring precise 3D data, enabling the creation of 3D models for intricate and expansive archaeological heritage sites. The data collection procedure for these sites frequently necessitates conducting multiple scans of both the interior and exterior from various perspectives in order to produce occlusion-free and lifelike 3D models. One potential approach involves employing a vehicle-mounted system to traverse the sites and gather exterior data, while wearable or handheld devices can be utilized to scan the interiors.

6.9 SUMMARY

In summary, the incorporation of intelligent sensors in the domains of localization, navigation, and mapping signifies a substantial shift in paradigm, leading to enhanced efficiency, precision, and dependability of these systems. The progress in sensor technology has facilitated unparalleled capabilities in self-governing operation, safety improvement, and resource efficiency across diverse sectors. Researchers and practitioners are continuously pushing the limits of localization accuracy, navigation precision, and mapping detail by harnessing the capabilities of intelligent sensors. The collection of location information plays a crucial role in the realization of the smart world. The reliance of various elements of the developing smart world on localization science and technology has been elucidated through an examination of diverse intelligent applications required for widespread implementation in smart devices, robots, smart health monitoring and delivery systems, smart spaces utilizing RFID tags, and emerging smart city transportation systems. It has been contended that the field of localization science and technology is a multifaceted phenomenon that necessitates additional investigation in order to meet the required levels of accuracy for these applications across various application settings and platforms. With the continuous advancement of smart sensor technologies and the emergence of new inventions, there is a great potential to revolutionize several applications, including transportation and infrastructure development. The findings presented in this chapter establish a fundamental basis for future research and development endeavors focused on maximizing the whole capabilities of intelligent sensors in revolutionizing our perception and engagement with the surrounding environment.

6.9.1 Future prospects

The current explosion of deep-learning applications and the big data market can be attributed in large part to geospatial intelligence. Many groups are beginning to see the value in mobile mapping solutions that work across platforms. The pioneers in the geospatial community who developed and advocate for mobile mapping technologies thirty years ago and their widespread use across diverse communities, including robotics, computer vision, AI, and navigation, has exceeded their initial expectations. First, the capacity to accomplish seamless mapping scenarios; second, the increasing use of cost-effective direct georeferencing devices; third, the increasing integration of AI; and fourth, the increasing adoption of unmanned multi-platforms for collaborative mapping are the four main markers of the future technological trends in mobile mapping that will improve urban informatics applications. Various mapping sensors attached to various unmanned and manned platforms, such as land vehicles, helicopters, and airplanes, are used for the gathering of geospatial data. Because of their ability to supply accurate and up-to-date geospatial data, mobile mapping systems play an indisputable role in urban informatics applications. This data is crucial for building the digital infrastructure that underpins urban informatics.

REFERENCES

Abdellatif, A. A., Mohamed, A., Chiasserini, C. F., Tlili, M., & Erbad, A. (2019). Edge computing for smart health: Context-aware approaches, opportunities, and challenges. *IEEE Network*, *33*(3), 196–203.

Adao, T., Magalhães, L., Fernandes, H., Paredes, H., & Barroso, J. (2012, June). Navigation module of Blavigator prototype. In: *World Automation Congress 2012* (pp. 1–6). IEEE, Puerto Vallarta.

Ahirwar, S., Swarnkar, R., Bhukya, S., & Namwade, G. (2019). Application of drone in agriculture. *International Journal of Current Microbiology and Applied Sciences*, *8*(1), 2500–2505.

Alanezi, M. A., Bouchekara, H. R., & Javaid, M. S. (2021). Range-based localization of a wireless sensor network for internet of things using received signal strength indicator and the most valuable player algorithm. *Technologies*, *9*(2), 42.

Alonso, M., Amaris, H., Alcala, D., & Florezr, D. M. (2020). Smart sensors for smart grid reliability. *Sensors*, *20*(8), 2187.

Braun, A., Wichert, R., Kuijper, A., & Fellner, D. W. (2015). Capacitive proximity sensing in smart environments. *Journal of Ambient Intelligence and Smart Environments*, 7(4), 483–510.

Buffi, A., Bernardini, F., Nepa, P., Marracci, M., Tellini, B., Di Donato, L., ... & Ferraro, A. (2021, September). RFID-based localization enables a smart system for worker safety. In: *2021 IEEE 6th International Forum on Research and Technology for Society and Industry (RTSI)* (pp. 513–518). IEEE, Naples.

Čech, M., Beltman, A. J., & Ozols, K. (2022, September). Digital twins and AI in smart motion control applications. In: *2022 IEEE 27th International Conference on Emerging Technologies and Factory Automation (ETFA)* (pp. 1–7). IEEE, Stuttgart.

Červenka, P., Břinda, K., Hanousková, M., Hofman, P., & Seifert, R. (2016). Blind friendly maps: Tactile maps for the blind as a part of the public map portal (Mapy. cz). In: *Computers Helping People with Special Needs: 15th International Conference, ICCHP 2016, Proceedings*, Linz, Austria, Part II, 15 July 13–15, 2016 (pp. 131–138). Springer.

Cho, B. G., Koo, J. K., Yoon, B. J., & Kim, B. W. (2010, October). The research of dead reckoning stabilization algorithm using different kinds of sensors. In: *ICCAS 2010* (pp. 1089–1092). IEEE, Gyeonggi-do.

Coronato, A., & Esposito, M. (2008, March). Towards an implementation of smart hospital: A localization system for mobile users and devices. In: *2008 Sixth Annual IEEE International Conference on Pervasive Computing and Communications (PerCom)* (pp. 715–719). IEEE, Hong Kong.

Ehlers, M. H., Huber, R., & Finger, R. (2021). Agricultural policy in the era of digitalisation. *Food Policy*, *100*, 102019.

Falcone, M., Hahn, J., & Burger, T. (2017). Galileo. In: P. J. G. Teunissen & O. Montenbruck (eds.), *Handbook of Global Navigation Satellite Systems* (pp. 247–272). Springer.

Fassi, F., & Perfetti, L. (2019). Backpack mobile mapping solution for DTM extraction of large inaccessible spaces. *International Archives of the Photogrammetry, Remote Sensing and Spatial Information Sciences*, *42*(2), 473–480.

Gu, Y., & Ren, F. (2015). Energy-efficient indoor localization of smart hand-held devices using Bluetooth. *IEEE Access*, *3*, 1450–1461.

Halvorsen, K. (2021). *Analysis of Position Estimation in a Dead Reckoning Navigation Robot* (Master's thesis, NTNU).

Indri, M., Lachello, L., Lazzero, I., Sibona, F., & Trapani, S. (2019). Smart sensors applications for a new paradigm of a production line. *Sensors*, *19*(3), 650.

Isazade, V. (2023). Advancement in navigation technologies and their potential for the visually impaired: a comprehensive review. *Spatial Information Research*, *31*(5), 547–558.

Jha, H., Lodhi, V., & Chakravarty, D. (2019, March). Object detection and identification using vision and radar data fusion system for ground-based navigation. In: *2019 6th International Conference on Signal Processing and Integrated Networks (SPIN)* (pp. 590–593). IEEE, Noida.

Jirawimut, R., Ptasinski, P., Garaj, V., Cecelja, F., & Balachandran, W. (2003). A method for dead reckoning parameter correction in pedestrian navigation system. *IEEE Transactions on Instrumentation and Measurement*, *52*(1), 209–215.

Khelifi, F., Bradai, A., Benslimane, A., Rawat, P., & Atri, M. (2019). A survey of localization systems in internet of things. *Mobile Networks and Applications*, *24*, 761–785.

Khoshelham, K., & Zlatanova, S. (2016). Sensors for indoor mapping and navigation. *Sensors*, *16*(5), 655.

Laituri, M., & Kodrich, K. (2008). On line disaster response community: People as sensors of high magnitude disasters using internet GIS. *Sensors*, *8*(5), 3037–3055.

Lauterbach, H. A., Borrmann, D., Heß, R., Eck, D., Schilling, K., & Nüchter, A. (2015). Evaluation of a backpack-mounted 3D mobile scanning system. *Remote Sensing*, *7*(10), 13753–13781.

Lee, J. H., & Hashimoto, H. (2003). Controlling mobile robots in distributed intelligent sensor network. *IEEE Transactions on Industrial Electronics*, *50*(5), 890–902.

Li, S., Zhang, M., Ji, Y., Zhang, Z., Cao, R., Chen, B., ... & Yin, Y. (2021). Agricultural machinery GNSS/IMU-integrated navigation based on fuzzy adaptive finite impulse response Kalman filtering algorithm. *Computers and Electronics in Agriculture*, *191*, 106524.

Liang, G., Ca, J., Liu, X., & Liang, J. (2016). Smart world: a better world. *Science China Information Sciences*, *59*, 1–5.

Lima, G. C., Figueiredo, F. L., Barbieri, A. E., & Seki, J. (2021). Agro 4.0: Enabling agriculture digital transformation through IoT. *Revista Ciência Agronômica*, *51*, e20207771.

Lovas, T., Hadzijanisz, K., Papp, V., & Somogyi, A. J. (2020). Indoor building survey assessment. *The International Archives of the Photogrammetry, Remote Sensing and Spatial Information Sciences*, *43*, 251–257.

Malavazi, F. B., Guyonneau, R., Fasquel, J. B., Lagrange, S., & Mercier, F. (2018). LiDAR-only based navigation algorithm for an autonomous agricultural robot. *Computers and Electronics in Agriculture*, *154*, 71–79.

Malinverni, E. S., Pierdicca, R., Bozzi, C. A., & Bartolucci, D. (2018, October). Evaluating a SLAM-based mobile mapping system: a methodological comparison for 3D heritage scene real-time reconstruction. In: *2018 Metrology for Archaeology and Cultural Heritage (MetroArchaeo)* (pp. 265–270). IEEE, Cassino.

Margiani, T., Cortesi, S., Keller, M., Vogt, C., Polonelli, T., & Magno, M. (2023). Angle of arrival and centimeter distance estimation on a smart UWB sensor node. *IEEE Transactions on Instrumentation and Measurement*.

Messaoudi, M. D., Menelas, B. A. J., & Mcheick, H. (2022). Review of navigation assistive tools and technologies for the visually impaired. *Sensors*, *22*(20), 7888.

Mittal, A., Tiku, S., & Pasricha, S. (2018, May). Adapting convolutional neural networks for indoor localization with smart mobile devices. In: *Proceedings of the 2018 on Great Lakes Symposium on VLSI* (pp. 117–122). Chicago, IL.

Mogili, U. R., & Deepak, B. B. V. L. (2018). Review on application of drone systems in precision agriculture. *Procedia Computer Science*, *133*, 502–509.

Moreno, M. V., Zamora, M. A., Santa, J., & Skarmeta, A. F. (2012, July). An indoor localization mechanism based on RFID and IR data in ambient intelligent environments. In: *2012 Sixth International Conference on Innovative Mobile and Internet Services in Ubiquitous Computing* (pp. 805–810). IEEE, Palermo.

Otero, R., Lagüela, S., Garrido, I., & Arias, P. (2020). Mobile indoor mapping technologies: A review. *Automation in Construction*, *120*, 103399.

Pahlavan, K., Krishnamurthy, P., & Geng, Y. (2015). Localization challenges for the emergence of the smart world. *IEEE Access*, *3*, 3058–3067.

Panigrahi, P. K., & Bisoy, S. K. (2022). Localization strategies for autonomous mobile robots: A review. *Journal of King Saud University-Computer and Information Sciences*, *34*(8), 6019–6039.

Paolini, G., Masotti, D., Antoniazzi, F., Cinotti, T. S., & Costanzo, A. (2019). Fall detection and 3-D indoor localization by a custom RFID reader embedded in a smart e-health platform. *IEEE Transactions on Microwave Theory and Techniques*, *67*(12), 5329–5339.

Patrik, A., Utama, G., Gunawan, A. A. S., Chowanda, A., Suroso, J. S., Shofiyanti, R., & Budiharto, W. (2019). GNSS-based navigation systems of autonomous drone for delivering items. *Journal of Big Data*, *6*, 1–14.

Patwari, N., Ash, J. N., Kyperountas, S., Hero, A. O., Moses, R. L., & Correal, N. S. (2005). Locating the nodes: cooperative localization in wireless sensor networks. *IEEE Signal Processing Magazine*, 22(4), 54–69.

Picinali, L., Afonso, A., Denis, M., & Katz, B. F. (2014). Exploration of architectural spaces by blind people using auditory virtual reality for the construction of spatial knowledge. *International Journal of Human-Computer Studies*, 72(4), 393–407.

Rajavarathan, J., Retscher, G., & Karunanithy, G. (2024). Implementation of GAGAN augmentation on smart mobile devices and development of a cooperative positioning architecture. *Journal of Applied Geodesy*, *18*(3), 541–552.

Revnivykh, S., Bolkunov, A., Serdyukov, A., & Montenbruck, O. (2017). GLONASS. In: P. Teunissen & O. Montenbruck (eds.), Handbook of Global Navigation Satellite Systems (pp. 219–245). Springer.

Saiz-Rubio, V., & Rovira-Más, F. (2020). From smart farming towards agriculture 5.0: A review on crop data management. *Agronomy*, *10*(2), 207.

Sehrawat, D., & Gill, N. S. (2019, April). Smart sensors: Analysis of different types of IoT sensors. In: *2019 3rd International Conference on Trends in Electronics and Informatics (ICOEI)* (pp. 523–528). IEEE, Tirunelveli.

Shen, S. (2013). Accurate multiple view 3d reconstruction using patch-based stereo for large-scale scenes. *IEEE Transactions on Image Processing*, 22(5), 1901–1914.

Shit, R. C., Sharma, S., Puthal, D., James, P., Pradhan, B., Van Moorsel, A., ... & Ranjan, R. (2019). Ubiquitous localization (UbiLoc): A survey and taxonomy on device free localization for smart world. *IEEE Communications Surveys & Tutorials*, *21*(4), 3532–3564.

Sparrow, R., & Howard, M. (2021). Robots in agriculture: prospects, impacts, ethics, and policy. *Precision Agriculture*, *22*, 818–833.

Spencer Jr, B. F., Ruiz-Sandoval, M. E., & Kurata, N. (2004). Smart sensing technology: opportunities and challenges. *Structural Control and Health Monitoring*, *11*(4), 349–368.

Stricker, R., Eisenbach, M., Sesselmann, M., Debes, K., & Gross, H. M. (2019, July). Improving visual road condition assessment by extensive experiments on the extended gaps dataset. In: *2019 International Joint Conference on Neural Networks (IJCNN)* (pp. 1–8). IEEE, Budapest.

Sugiyama, W. (2020). Voice guidance in Maps, built for people with impaired vision. https://www.blog.google/products/maps/better-maps-for-people-with-vision-impairments/

Veroustraete, F. (2015). The rise of the drones in agriculture. *EC Agriculture*, *2*(2), 325–327.

Wang, W. Q. (2016). Overview of frequency diverse array in radar and navigation applications. *IET Radar, Sonar & Navigation*, *10*(6), 1001–1012.

Wiki, O. (2020). Precision of coordinates–openstreetmap wiki.

Winter, S., Tomko, M., Vasardani, M., Richter, K. F., Khoshelham, K., & Kalantari, M. (2019). Infrastructure-independent indoor localization and navigation. *ACM Computing Surveys (CSUR)*, *52*(3), 1–24.

Zhang, Y., Gu, Y., Vlatkovic, V., & Wang, X. (2004, June). Progress of smart sensor and smart sensor networks. In: *Fifth World Congress on Intelligent Control and Automation (IEEE Cat. No. 04EX788)* (Vol. 4, pp. 3600–3606). IEEE, Hangzhou.

Zhao, Y. (2000). Mobile phone location determination and its impact on intelligent transportation systems. *IEEE Transactions on Intelligent Transportation Systems*, *1*(1), 55–64.

Chapter 7

Structural health monitoring system

Vibration analysis using a ZigBee module

Ashwini Jatti, Shubhangi Deokar, Javed Sayyad, Anup Vibhute, and Lalit Chaudhari

7.1 INTRODUCTION

Civil infrastructure, including bridges, premises, pipelines, and offshore structures, is crucial for human life and disaster management. Structural health monitoring (SHM) is a developing area that uses sensors to assess and repair these structures, minimizing the effects of disasters and aiding in society's recovery. SHM (Haque et al. 2015) should use less energy and function as an early warning system for structural problems. Two distinct sensing modules, flex detecting and acceleration sensing, are used for vibration analysis. Understanding ground and structure vibration during an earthquake, earthquake resistance design, and building response is necessary. Real-time monitoring algorithms with damage detection algorithms are needed to analyze seismic reactions effectively. The intelligent environment, the next evolution in infrastructure automation, relies on sensory information from the outside world. Distributed wireless sensor networks (WSNs) supply the data required for smart environments, consisting of a data distribution network and a data gathering network monitored and managed by a management center.

The advancement of SHM has become a significant technology in improving the reliability and safety of aeronautical and highly stressed civil structures, while also supporting their efficient maintenance. In the field of structural engineering, wireless networks are now considered as an effective alternative to traditional tethered control systems. Implementing wireless structural monitoring systems is more cost-effective due to the elimination of extensive wiring between sensors and data acquisition systems. To evaluate the efficacy of this proposed approach for measuring vibrations in large-scale civil buildings, we have developed a wireless modular monitoring system that utilizes timber flooring supported by aluminum pillars. By utilizing an accelerometer, flex sensor, and MATLAB® Simulink software, this system is capable of generating an alarm when vibrations exceed a predetermined threshold level. Furthermore, it can send an emergency message to registered mobile devices via the global system for mobile

DOI: 10.1201/9781003633884-7

communications (GSM) module. Communication between the client and server occurs wirelessly through ZigBee technology.

This chapter an overview of the requirements for WSNs. This involves doing away with the current practice of most wireless systems requiring battery replacements. It makes sense to presume that a monitoring system's price should not significantly add to a building's overall cost. Energy problems continue to impede monitoring apps. The savings from wireless sensors would quickly outweigh their costs because batteries are costly to replace. The disadvantages of wired instrumentation and wired energy distribution are the same. Although the power levels are very low, it is feasible to harvest energy from the surrounding environment without ever running out of power. Sensor energy usage must be kept to a minimum for energy harvesting technologies.

7.1.1 Challenges

In civil infrastructure applications (Heo and Jeon 2009), wireless monitoring systems play a crucial role in analyzing and processing data at a local level. These systems are expected to have the ability to communicate over relatively long distances within the structure span, which can range from tens of meters to hundreds of meters. However, potential end users need to consider the power consumption of wireless sensing units as they typically operate on batteries with limited power.

Designing a wireless structural sensing and monitoring system involves finding a balance between the demand for higher data processing and communication capacity, while also minimizing power consumption. This is because long-distance communication and local data interrogation generally require more power. Additionally, ensuring reliable and accurate data acquisition poses significant challenges for these systems.

To achieve this, sensor signals are converted from analog to digital format before being temporarily stored. Wireless communication is then utilized to transfer the digitized signals. It is essential to appropriately handle circuit noise and failures in wireless communication to ensure precise and dependable data acquisition.

7.2 PREVIOUS WORK ON DEVELOPING STRUCTURAL HEALTH MONITORING

A wireless sensing unit for SHM has been designed by Wang et al. (2007) and Cho et al. (2008). A powerful reduced instruction set computer (RISC) microcontroller is employed in the sensing unit to process and aggregate data. Two micro-electro-mechanical systems (MEMS) based accelerometers were additionally applied in this research. Using global positioning system (GPS), Knecht and Manetti (2001) presented SHM; positioning data can be provided within millimeters in near real time. Differential carrier phase GPS can monitor and oversee small movements utilizing a network of receiver modules. Landslides and rock formations can also be measured with this kind of sensor network in addition to structures such as buildings, dams, and bridges. Wang et al. (2007) and Cho et al. (2008) verified wireless modular monitoring devices for structures.

The design and construction of a device for a wireless modular monitoring system (WiMMS). The wireless sensing unit provides a low-cost, high-performance option for tracking structural performance over the short and long term by combining cutting-edge wireless transmission technology, low-power microprocessors, and MEMS sensing transducers. In this research, we explored the idea of using the computational core of the WSN units to query local data. Fast Fourier transform implementation using wireless sensors mounted on test structures was effective, and the primary response modes of the structure were discovered.

A method for monitoring bridge health under ordinary traffic loads was presented by Lee and Shinozuka (2006) and Knecht and Manetti (2001). Using ambient vibration data brought on by traffic loadings, a technique for calculating bridge damage is presented. Damage locations and severity are assessed based on the operational modal properties. The modal properties of vibrations caused by ordinary traffic loads are used to determine the element-level damage to a bridge. Vehicle tests on a bridge model are used to verify the method. Under a variety of vehicle load circumstances, the majority of the inflicted damages can be identified with success; however, the severity of the damage is frequently overestimated. Wang et al. verified real-time wireless monitoring of civil structures using an integrated network system (Wang et al. 2007). The suggested system supports several wireless sensing units, enabling real-time data acquisition and analysis from a variety of analog sensors. The viability and dependability of the system have been validated by numerous laboratory as well as field experiments (Kim et al. 2007a).

Farrar et al. conducted a coupling of sensing hardware and data interrogation software (Farrar et al. 2006; Mascarenas et al. 2007; Overly et al. 2008). They have tackled the SHM issue as a component of a statistical pattern recognition paradigm. With the help of this paradigm, a four-step process can be explained: operational assessment, data collection and cleaning, feature extraction and compression, and statistical model development for feature discrimination. These procedures are typically implemented using a combination of hardware and software. Kang et al.'s investigation into structural system identification (SI) in the time domain to calculate stiffness and damping factors used measured acceleration (Kang et al. 2005). The least squares errors between the detected acceleration and the calculated acceleration by an arithmetical model of a structure are known as error functions. The effectiveness of the suggested technique is shown in an experiment using a three-story shear building model and a numerical simulation study.

The effectiveness of a WSN for SHM was researched by Paek et al. (2005). WISDEN, a wireless multi-hop sensor network-based data acquisition system for SHM applications, was assessed for performance by the authors. Both an earthquake test structure and a four-story office building served as the real-world deployment sites for WISDEN. During this experiment, WISDEN's performance fell short of expectations, mainly because of a software bug. Studies by Lu et al. (2005) examined the application of wireless sensors for SHM tracking and management. A three-story half-scale steel structure was fitted with a wireless monitoring system made up of six wireless sensors to evaluate the performance of WiMMS for measuring vibration in large-scale civil structures. Data communication reliability was confirmed using a shaking table. Based on the findings, WiMMS can be applied for a variety of monitoring and control uses related to civil infrastructure.

Using a wireless sensing unit, the converter can be used to observe various sensor signals. For structural control, it is possible to integrate the control algorithm and control gain into the sensing device. For a real-time, nondestructive assessment of steel bridge members, Park et al. (2006a,b) used active sensing. Their research explores the potential for nondestructive evaluations of steel bridge members using active sensing methods based on piezoelectric lead-zirconate-titanate (PZT). Impedance-based damage detection and lamb wave-based damage detection were both explored as damage detection techniques for steel bridge members. The effectiveness of the suggested techniques was supported by experimental findings.

The research by Chung et al. (2004) concentrated on the wireless MEMS for real-time structural response visualization. The purpose of this study was to create and test wireless MEMS-type sensors for real-time earthquake monitoring on bridges. Without the need for multiple cables, MEMS sensors with wireless capabilities can be used for bridge monitoring. Possible damage to bridge structures can be discovered by further developing the embedded microcomputer units and functions, such as detecting frequency responses, to provide early notice to citizens and stop the loss of life or property damage.

Kim et al. (2007b) outlined the use of WSNs to check on the condition of the civil infrastructure in their research. A WSN for SHM is installed on the main span and south tower of the Golden Gate Bridge and has been planned, developed, deployed, and tested. Ambient structural vibrations can be accurately measured at a reasonable expense and without hindering bridge operation. In order to satisfy the requirements that SHM places on WSN, solutions are proposed and put into practice. Raghavan and Cesnik (2007) have given a summary of guided-wave SHM.

Results of an international collaborative study project on smart wireless sensors and SHM of civil structures were presented by Chou et al. (2008) to the Korea Institute of Science and Technology, the University of Michigan, and the University of Illinois at Urbana Champaign. The state of the art in smart wireless sensor technology is discussed, along with a number of subsystems of a smart wireless sensor and wireless sensor platforms created both in academia and business. An infrastructure monitoring system was developed by Heo and Jeon (2009) using ubiquitous computing technology. Using a wireless sensor with a feature of Bluetooth technology, this system communicates data measured by the wireless sensing unit using transmission control protocol/internet protocol (TCP/IP) network protocol. The randomly excited self-anchored suspension bridge was tried for real-time network user monitoring. The monitoring test data analysis uncovered dynamic features. As opposed to using input excitation devices for structural monitoring, real-time tests were conducted under ambient vibrations.

WSNs have been used for SHM in historical buildings by Anastasi et al. (2009). A baroque building dating back to the early 1700s in Palermo, Italy, the church of St. Teresa, is the subject of the author's discussion of the use of WSNs on it. Grisso and Inman (2005) created an autonomous impedance-based SHM device in orbit for thermal protection. Early work has concentrated on creating a digital signal processor-based prototype in order to realize a completely active sensor system self-contained using impedance-based SHM. Structural excitation, data collection, and health monitoring analysis happen in just a few seconds, as opposed to conventional impedance methods, which demand that all analyses be completed using processing software after data acquisition.

Using impedance-based SHM techniques, Park et al. (2006a) identified multiple cracks in concrete buildings (Chung et al. 2004). The authors of this research offer a feasibility study on the use of PZT patches for real-time SHM on concrete structures. First, a comparison between experimental and analytical techniques for spotting damage to a simple concrete beam was performed. In an experimental investigation, PZT patches were used in both lateral and thickness modes to evaluate artificially induced progressive surface damage on plain concrete beams. The validity of the experimental findings was then confirmed through an analytical study based on finite element (FE) models. This research demonstrated successful damage prediction outcomes with a steady trend in the variation of the PZT patches' impedance signature. The impedance-based developed by Mascarenas et al. (2007) with a wireless sensor node was designed for SHM. High-frequency ultrasonic vibrations are used to excite the structure using surface-bonded piezoelectric transducers.

An integrated circuit chip equipped with a wireless impedance sensor node based on a piezoelectric transducer can be used to measure and record its electrical impedance in this study. In addition to the microcontroller, which performs local computation, the base station (BS) receives structural information via wireless telemetry. For impedance-based SHM, Kim et al. employed a system-on-board strategy. By bringing a new excitation technique and a new damage detection system, we are able to do away with the need for analog-to-digital and digital-to-analog conversions. A test structure was used to assess the prototype's performance and confirm its functionality. Based on measurements made with the test structure, the SHM prototype delivers consistent efficiency. Overly et al. (2008) created a remarkably small wireless impedance-detecting device as a component of the SHM and piezoelectric active-sensor self-diagnostics. A newly created, low-cost integrated circuit is used in the sensor node to measure and record the electrical impedance of a piezoelectric transducer. Lee and Shinozuka (2006) measured real-time displacements on a flexible bridge using digital image processing methods. This technique is very simple to use, extremely cost-effective, and maintains the benefits of dynamic measurement and high resolution. The applicability and efficacy of the present approach were evaluated using a steel-plate girder bridge and a steel-box girder bridge.

A vision-based displacement measurement method has been created and implemented in civil structures for SHM, according to Lee and Shinozuka (2006). The effectiveness of the suggested system was validated by contrasting the load-carrying capacities of the conventional sensor and the system under consideration. Comparison of the movement measured by the vision-based system with that measured by traditional contact-type sensors and linear variable differential transformers (LVDT) served as the basis for verification. Lamb waves can be used to find damage in composite materials, as reported by Kessler et al. (2002). This article presents in situ damage detection for composite materials as part of a practical and analytical review. The results of applying Lamb wave techniques to quasi-isotropic graphite/epoxy test specimens with representative damage modes are reported in this article.

Jang et al. (2002) carried out experimental research on SI-based damage assessment of structures. To measure damage assessment algorithms based on the SI method, laboratory tests are carried out. On a grid-style model bridge, laboratory tests were performed to quantify static displacements and modal information from impact vibrations. Experimental data were used to establish the standard structure and properties from the undamaged model structure. The structural stiffness around the crack was decreased as a consequence of the damage, which was simulated using different saw-cutting depths. The applicability of SI-based damage evaluation algorithms has been studied experimentally. Piezoelectric materials can harness the force of vibration, as reported by Sodano et al. (2004). Power harvesting is a technique that makes use of piezoelectric materials to harness the vibrations that naturally surround a device. Due to their crystalline structure, piezoelectric materials can turn mechanical strain energy into electrical charge and electrical charge into mechanical strain, or vice versa, depending on the applied electrical potential.

The difficulties posed by next-generation wireless sensing networks and their effects on society were covered by Srivastava (2010). The categorization of WSNs and the difficulties facing the upcoming WSN generation are covered by the author. In a handbook of Ilyas and Mahgoub (2004), discusses sensor networks, which consist of multiple electronic sensors that communicate and interact with compact wireless and wired sensor networks. Thongsopa et al. (2010) measured the strength of the ultra high frequency (UHF) signal that was propagating from the road surface in their research when there were obstructions from vehicles. It is challenging to use WSNs for traffic tracking due to radio-wave propagation on the road surface. A car is on an unoccupied road, but a car is not on an empty road. The receiving signal strength in these two cases is compared by the writers. To ascertain how antenna polarization and height affect received signal strength, a research is carried out. The transmission antenna was mounted on the asphalt of the road. The receiving signal was recorded 360° around the transmitting antenna using the 2.5-m radius antenna. Both situations exhibit fluctuating reception near the transmitter based on the observations. Signals are received by vertically polarized receivers.

Omnidirectional antenna ground effects in wireless sensing networks were investigated by Janek and Evans (2010). Omnidirectional antennas are frequently used in WSNs to perform radio frequency (RF) communication. Outside noise, electromagnetic interference, overloaded network traffic, significant barriers (vegetation and buildings), the composition of the topography and the atmosphere, and weather patterns can all cause signal degradation, the loss of data packets, and a reduction in the range of radio communications. The objective of this study is to examine the RF range reduction capabilities of a specific WSN created to operate in agricultural crop areas and gather aggregate soil moisture and soil temperature data. The study shows that when antennas are positioned close to the ground, omnidirectional transmission patterns greatly alter (within 10 cm).

7.3 PROPOSED METHODOLOGY

Various architectures are being studied using microprocessor specifications and power harvesting techniques. Due to complex architectural design, SHM methods for civil engineering structures can be designed for energy-efficient wireless sensing systems, monitoring structures automatically and in real time, and evaluating building safety.

Batteries are necessary for wireless systems to manage electricity. Each unit has a distinct set of power modules. In order to power all the modules, it is necessary to regularize the voltage at different steps. The sensing module and the power device must be integrated. The sensing module is made up of various kinds of sensors that are used to assess the various factors that have an impact on structural health.

The sensor module receives environmental signals and sends them to the acquisition module for power consumption, vibration analysis, and damage detection. The acquisition module includes a controller, signal conditioner, and ADC. Signal conditioning performs isolation, amplification, and filtration tasks. ZigBee-based wireless communications provide low power consumption, dependability, and affordability for these devices. In commercial and industrial uses, it is very simple to integrate with many products. Numerous energy management areas are well suited for the ZigBee-based communication technology. A WSN is a cutting-edge device, particularly for commercial and industrial uses.

7.3.1 Vibration detection method

An earthquake prediction module is shown in Figure 7.1. Essentially, the module predicts earthquakes before humans become aware of them. An earthquake can be detected using it. It is therefore necessary to identify vibrations caused by earthquakes. The most common cause of damage caused by earthquakes is vibrations in the ground. Sensors that measure vibrations use accelerometers and strain gauges.

Every level of the structure has an accelerometer, and the base has strain sensors. In earthquake prediction, accelerometers are mainly used in the early stages. When accelerometers detect vibrations at a level above the threshold, they can wirelessly transmit alert commands to the BS. An alert command will be sent to the BS when vibrations exceed 10–25 Hz. A flex sensor is used for measuring ground vibration in the same way as an acceleration sensing module. Using the flex sensor, you can mount a building on the base. According to the size of the building, different areas of the building are equipped with sensors. When ground vibrations exceed a threshold

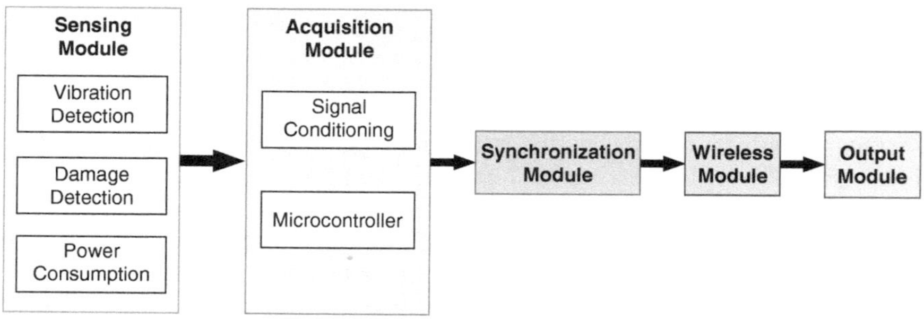

Figure 7.1 Conceptual block diagram.

level, ZigBee will alert the BS. Acceleration and strain can be detected using the earthquake detection system (Figure 7.1).

Accelerometers measure vibrations or accelerations. Several applications of accelerometers have been found in science and industry. It can be used to detect vibrations caused by outside forces like earthquakes and typhoons in civil engineering. Accelerometers, which are exemplary sensor types, can be used to detect structure vibrations. Depending on their location and intended use, three kinds of accelerometers can be used in proposed systems. In addition to force-balance accelerometers, there are also piezoelectric and MEMS accelerometers accessible. This proposed technique measured vibrations using a MEMS accelerometer. This kind of accelerometer works by suspending a small mass from tiny structures that are carved into silicon surfaces. Newton's second law states that any object that feels a force and a mass displacement accelerates or vibrates. Micro-electromechanical accelerometers are compact and have a minimal power requirement.

In addition to force-balance accelerometers, piezoelectric accelerometers and MEMS accelerometers are also available. In this proposed method, vibrations were measured using a MEMS accelerometer. To operate this type of accelerometer, a small mass is etched into silicon surfaces and suspended by small structures. Micro-electromechanical accelerometers, in addition to measuring frequencies up to 300 Hz, are also capable of measuring power. Any structure can use strain sensors to measure external strains. Electrical resistance sensors are the most used in real-time structures. Electrical resistance changes with mechanical deformation in a conductor, which is the basis of this technique. Strain causes it to be highly sensitive. It has an output voltage of 2.5 mV and can measure strains in the range of around 0–500 microstrains. The signal obtained from the accelerometer and strain sensor is processed using a peripheral interface controller (PIC) microcontroller of the PIC16F874A type. It can be a high-performance RISC central processing unit (CPU) yet only contains 35 basic programming instructions. It includes eight tiers of deep hardware stacks and a clock input frequency of DC-20 MHz for its operating speed. The primary function is sleep mode, which is power saving mode. It can resist a variety of industrial and commercial temperatures. Lastly, it results in little power usage.

7.3.2 Process flow for earthquake wake-up

1. Earthquake wake-up is detected in both vibrations and strains.
2. BS receives the warning message from an accelerometer and strain sensor when the maximum threshold level is surpassed by their output.
3. A BS takes control here to initiate the monitoring of nodes depending upon their capability of sending warning signals.

4. Monitoring nodes receive a wake-up indication from the BS. As a result, monitoring sites are dynamically and locally chosen. BSs can enable or disable monitoring nodes. All parameters are configured in the BS.
5. The monitoring node needs to be woken up by the BS when an external event occurs. The monitoring node needs to wake up in a limited amount of time to prevent the loss of data. The BS reads the data after the event has occurred.
6. A new event occurs, and the wake-up algorithm holds the data for the second event and finishes reading out the first event's data.
7. After that, the second event's data can be read out. The distributed earthquake detection system is shown in Figure 7.2.

7.3.3 Output module

A ZigBee module, a PC, a GSM module, and a mobile device are included in the output module. Messages are displayed on mobile devices and output is displayed through PCs, as shown in Figure 7.3.

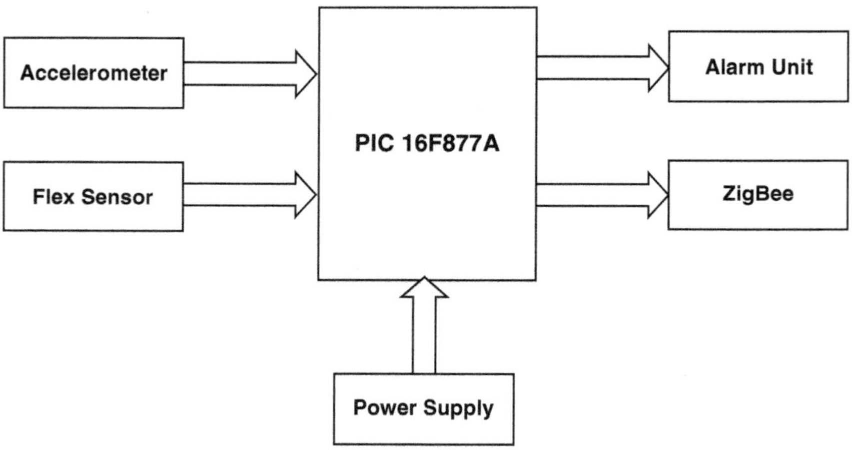

Figure 7.2 Earthquake detection system.

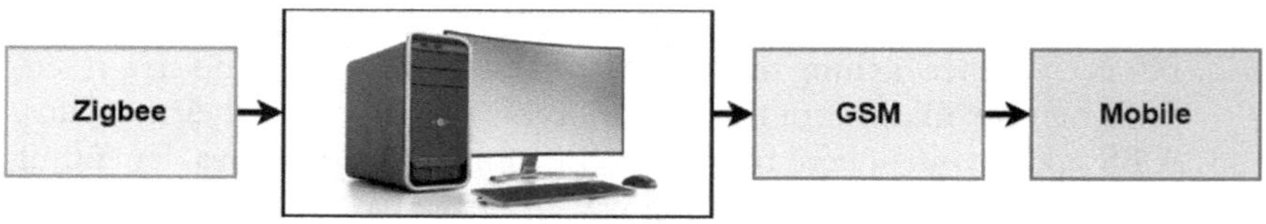

Figure 7.3 Block diagram of output module.

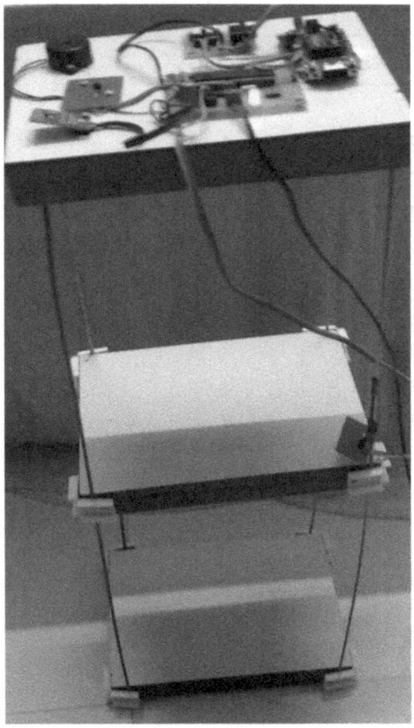

Figure 7.4 Prototype SHM system.

ZigBee is connected to the PC and receives data from input modules. The input module is assumed to be the client node in this method. The server node will be the output module. Client nodes wirelessly transmit data to server nodes via ZigBee.

The ZigBee pair that is employed by both nodes must correspond. Each module's result is shown on the computer. The early stages of an earthquake and an abnormal vibration are the first module's outputs. HyperTerminal is used to show these outputs on a computer. This device serves as a master device. A warning message is sent to the registered mobile via the GSM module through the server node when the vibration is above the threshold level. This section depicts the system integration and implementation of SHM on a prototype. Figure 7.4 shows the two-floor civil structure on which the SHM system is being mounted to capture the vibration using acceleration and flex sensors.

7.4 RESULTS

The accelerometer and flex sensor were calibrated before installing them in the wireless system module. Tables 7.1 and 7.2 show the calibration results for the accelerometer and flex sensor, respectively.

Table 7.1 Accelerometer calibration results

Angle, degrees	*Voltage (+X), V*	*Voltage (−X), V*	*Voltage (+Y), V*	*Voltage (−Y), V*
0	1.17	1.17	1.16	1.15
5	1.23	1.15	1.19	1.14
10	1.25	1.13	1.2	1.13
15	1.28	1.1	1.21	1.13
20	1.29	1.08	1.23	1.12
25	1.31	1.07	1.24	1.11
30	1.32	1.05	1.25	1.1
35	1.33	1.04	1.26	1.08
40	1.34	1.02	1.28	1.07
45	1.37	1	1.31	1.06

Table 7.2 Flex sensor calibration results

Angle, degrees	*Output, Ω*
50	185
60	186
70	189
80	192
90	205
100	208
110	213
120	217
130	220

To investigate the operating efficiency of the wireless SHM for free vibration test prototype structure is made up of a wooden floor with aluminum pillars. Table 7.1 shows the output results of the accelerometer and flex sensor before applying vibration. Figure 7.5 shows a picture of the collapsed building structure. When the vibration exceeds the set threshold voltage of 1.2 V, then there is a fluctuation in the output of the accelerometer and flex sensor observed, as shown in Tables 7.1 and 7.2. This leads to an alarm mounted on the building, and an emergency message is notified to the user's device via the GSM module through master nodes.

7.5 CONCLUSION

The purpose of the research was to provide insight into the utilization of embedded systems, presenting a novel controlling system that is both cost-effective and versatile for managing structural life. The architecture aims to simplify

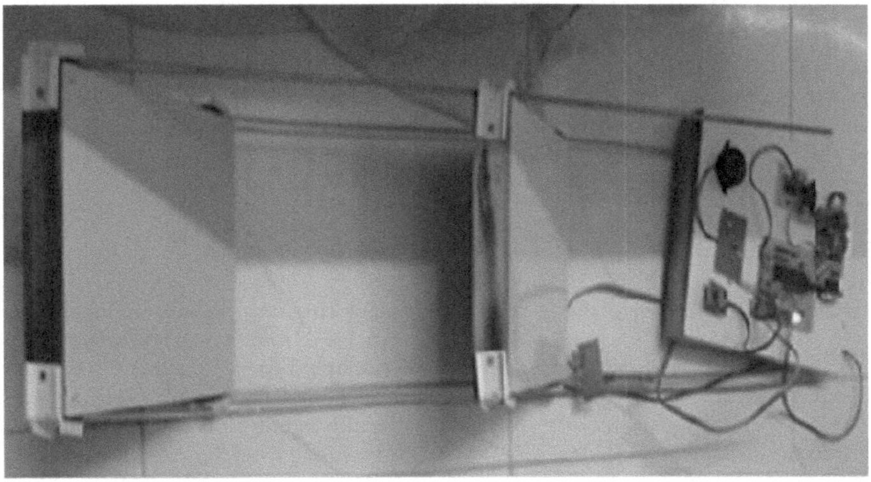

(a)

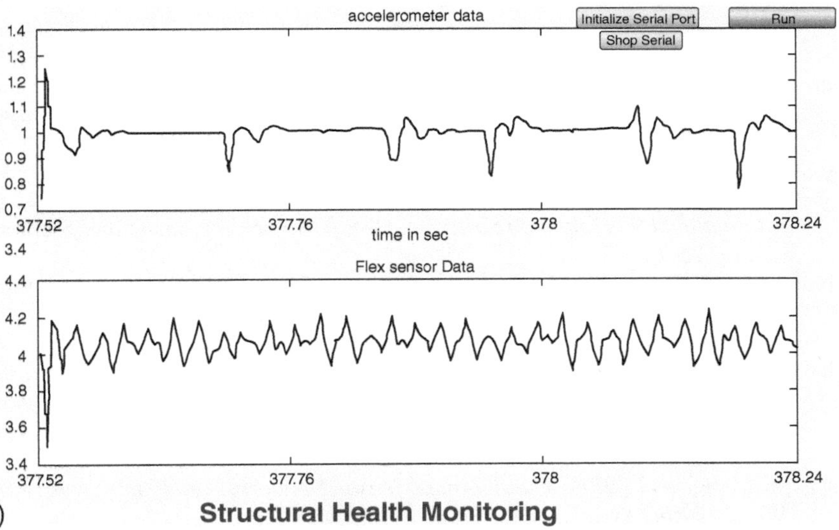

(b)

Figure 7.5 (a) Collapsed building and (b) system output graph after vibration.

systems and reduce the amount of physical wiring, thereby avoiding the use of complex and expensive components. To evaluate the effectiveness of a wireless-based monitoring system for assessing vibrations in large-scale infrastructure premises, a wireless modular monitoring system was installed on the top floor of a two-story half-scale structure. This device is equipped with an alarm that activates when vibration levels exceed a certain threshold, while also sending an emergency notification to registered mobile devices via a GSM module. The method employed is flexible and scalable, allowing for easy addition of more devices without much effort. Users have control over linked devices within their laboratory settings. By utilizing ZigBee technology, operating costs are minimized in this system. Furthermore, it provides online access and features a user-friendly interface that can be easily customized as needed. Additional experiments and improvements can be implemented to further refine this system's functionality.

REFERENCES

Anastasi G, RG Lo, Ortolani M (2009) WSNs for structural health monitoring of historical buildings. In: *2009 2nd Conference on Human System Interactions*, Catania, pp. 574–579. https://ieeexplore.ieee.org/abstract/document/5091041

Cho S, Yun CB, Lynch JP, Zimmerman AT, Spencer Jr, BF, Nagayama T (2008) Smart wireless sensor technology for structural health monitoring of civil structures. *Steel structures*, 8(4), 267–275.

Chung H-C, Enomoto T, Shinozuka M et al. (2004) Real time visualization of structural response with wireless MEMS sensors. In: *Proceedings of 13th World Conference on Earthquake Engineering*, August 1–6, 2004, Vancouver, BC, vol. 121, pp. 1–10.

Farrar CR, Allen DW, Park G et al. (2006) Coupling sensing hardware with data interrogation software for structural health monitoring. *Shock and Vibration* 13:519–530.

Grisso BL, Inman DJ (2005) Developing an autonomous on-orbit impedance-based SHM system for thermal protection systems. In: *Proceedings of the 5th International Workshop on Structural Health Monitoring*, September, Stanford, CA, pp 12–14.

Haque ME, Zain MFM, Hannan MA, Rahman MH (2015) Building structural health monitoring using dense and sparse topology wireless sensor network. *Smart Structures and Systems* 16:607–621.

Heo G, Jeon J (2009) A smart monitoring system based on ubiquitous computing technique for infra-structural system: Centering on identification of dynamic characteristics of self-anchored suspension bridge. *KSCE Journal of Civil Engineering* 13:333–337.

Ilyas M, Mahgoub I (2004) *Handbook of Sensor Networks: Compact Wireless and Wired Sensing Systems*. CRC Press.

Janek JF, Evans JJ (2010) Predicting ground effects of omnidirectional antennas in wireless sensor networks. *Wireless Sensor Network* 2:879.

Jang J-H, Yeo I, Shin S, Chang S-P (2002) Experimental investigation of system-identification-based damage assessment on structures. *Journal of Structural Engineering* 128:673–682.

Kang JS, Park S-K, Shin S, Lee HS (2005) Structural system identification in time domain using measured acceleration. *Journal of Sound Vibration* 288:215–234.

Kessler SS, Spearing SM, Soutis C (2002) Damage detection in composite materials using Lamb wave methods. *Smart Materials and Structures* 11:269.

Kim J, Grisso BL, Ha DS, Inman DJ (2007a) A system-on-board approach for impedance-based structural health monitoring. In: *Sensors and Smart Structures*

Technologies for Civil, Mechanical, and Aerospace Systems, edited by Masayoshi Tomizuka, Chung-Bang Yun, Victor Giurgiutiu, pp 244–252. SPIE. https://doi.org/10.1117/12.715791

Kim S, Pakzad S, Culler D et al. (2007b) Health monitoring of civil infrastructures using wireless sensor networks. In: *Proceedings of the 6th International Conference on Information Processing in Sensor Networks*, Cambridge, MA, pp. 254–263. https://dl.acm.org/doi/proceedings/10.1145/1236360

Knecht A, Manetti L (2001) Using GPS in structural health monitoring. In: *Smart Structures and Materials 2001: Sensory Phenomena and Measurement Instrumentation for Smart Structures and Materials*, SPIE's 8th Annual International Symposium on Smart Structures and Materials, 2001, Newport Beach, CA, vol. 4328, pp 122–129. SPIE. https://www.spiedigitallibrary.org/conference-proceedings-of-spie/4328/0000/Using-GPS-in-structural-health-monitoring/10.1117/12.435515.short

Lee J-J, Shinozuka M (2006) Real-time displacement measurement of a flexible bridge using digital image processing techniques. *Experimental Mechanics* 46:105–114.

Lu KC, Wang Y, Lynch JP et al. (2005) Application of wireless sensors for structural health monitoring and control. In: *Proceedings of KKCNN Symposium on Civil Engineering*, Taiwan.

Mascarenas DL, Todd MD, Park G, Farrar CR (2007) Development of an impedance-based wireless sensor node for structural health monitoring. *Smart Materials and Structures* 16:2137.

Overly TGS, Park G, Farinholt KM, Farrar CR (2008) Development of an extremely compact impedance-based wireless sensing device. *Smart Materials and Structures* 17:65011.

Paek J, Chintalapudi K, Govindan R, Caffrey J, Masri S (2005, May). A wireless sensor network for structural health monitoring: Performance and experience. In *The Second IEEE Workshop on Embedded Networked Sensors*, 2005. EmNetS-II, pp. 1–9. IEEE.

Park S, Ahmad S, Yun C-B, Roh Y (2006a) Multiple crack detection of concrete structures using impedance-based structural health monitoring techniques. *Experimental Mechanics* 46:609–618.

Park S, Yun C-B, Roh Y (2006b) Active sensing-based real-time nondestructive evaluations for steel bridge members. *KSCE Journal of Civil Engineering* 10:33–39.

Raghavan A, Cesnik CES (2007) Guided-wave signal processing using chirplet matching pursuits and mode correlation for structural health monitoring. *Smart Materials and Structures* 16:355.

Sodano HA, Inman DJ, Park G (2004) A review of power harvesting from vibration using piezoelectric materials. *Shock and Vibration Digest* 36:197–206.

Srivastava N (2010) Challenges of next-generation wireless sensor networks and its impact on society. *Journal of Telecommunications* 1(1): 128–133.

Thongsopa C, Sukphongchirakul P, Intarapanich A, Jarataku P (2010) Measurement of UHF signal strength propagating from road surface with vehicle obstruction. *World Academy of Science, Engineering and Technology* 45:123–127.

Wang Y, Swartz RA, Lynch JP et al. (2007) Decentralized civil structural control using real-time wireless sensing and embedded computing. *Smart Materials and Structures* 3:321–340.

Chapter 8

Wireless EEG recording and ANN classifier for epileptic classification using particle swarm optimization

Arun Kumar Wadhwani, Manoj Kumar Bandil, Bhavya Shukla, and Harsh Wadhwani

8.1 INTRODUCTION: BACKGROUND AND DRIVING FORCES

A seizure is a bursting event because of rare, intense, hyper-synchronous releases by accumulated neurons in the central nervous system. In childhood age, seizures are likely to occur extremely. The person faces frequent seizures due to a perpetual progression during neurological conditions like epilepsy. In order to detect epilepsy, the electroencephalogram (EEG) is very useful. Two types of activity can be observed in the EEG recordings of the epileptic patients: inter-ictal (EEG recordings show anomalies such as isolated spikes, sharp waves, and wave complexes) and ictal (EEG recordings disclose abnormity of the continuous dump, including isolated spikes, sharp waves, and wave complexes over time and usually clinically correlated).

In general, doctors have identified cases of epilepsy through visual examination of EEG in the ictal state. However, there is a serious drawback to the visual evaluation of huge EEG results. Visual EEG analysis, particularly in patient cases of long-duration EEG recording, is inefficient and time-consuming. In addition, because of the subjective study and variety of inter-ictal spikes morphology, the opinions and findings of the neurophysiologist might differ for the same EEG recording [1]. These drawbacks leads to the need for automated detection methods for inter-ictal spike events and the incidence of epileptic seizures to provide valuable clinical testing resources for competent epileptic diagnosis [2]. EEG range is typically split into five substrates: the delta substrip (0–4 Hz), theta substrip (4–8 Hz), alpha substrip (8–12 Hz), beta substrip (13–30 Hz), and gamma substrip (30–60 Hz). Since the entire EEG range is not successful compared with the individual frequency sub-bands in terms of brain dynamic representations, it reflects more detailed information on the related neuronal activity behind the signal. In the original EEG signal range, affected behaviors of EEG signals are not evident but can be increased by the study of each sub-band individually. The statistical and morphological characteristics [3]

DOI: 10.1201/9781003633884-8

are determined separately for EEG signal datasets of three different classes without wavelet decomposition and with wavelet decomposition, taking account of 1D information coefficients. This technique is used to study on the datasets consisting of three distinct EEG signals: (1) normal person, (2) seizure-free patients (inter-ictal or pre-seizure), and (3) seizure-free (ictal) [4]. Following the extraction of several features, it is our objective to boost pattern classification accuracy with and without implementing PSOs for selecting the no-cached-layer ANN (multilayer back propagating) grade by evaluating training efficiency and grade accuracies with and without time-varying acceleration coefficient (TVAC).

8.2 WAVELET-BASED ECG SUB-BAND DECOMPOSITION

In contrast to conventional methods like Fourier transform, wavelet transform is used to extract EEG signal sub-bands (delta, theta, alpha, beta, and gamma) because it has advantages of time-frequency localization, scale-space processing, and multi-resolution filtering [5]. Wavelet transform can be compressed and distributed by signal frequency, as this method covers the length of the signal [5,6] with variable window sizes. This method is useful for extracting features for multi-resolution analysis in nonstationary signals such as EEGs. In this work, decomposition with mother wavelet db4 up to the next stage and results of the multi-resolution study are given by five EEG sub-bands [7]. The EEG sub-bands undergo the consequent extraction of features.

8.3 METHODOLOGY

DWT supports to investigate the signal by multi-resolution process at different frequency bands by decomposing the signal into approximation and detail coefficients. The strategy for frequency band partition is executed in MATLAB® (2013a). Epilepsy discovery utilizing EEG requires highlight extraction from the obtained signal during explicit frequency scope of delta, theta, alpha, beta, and gamma sub-bands. This examination unambiguously depicts strategy for up-testing and recombining of deteriorated sub-bands to accomplish the necessary frequencies. Informational index signals have pre-prepared by evacuating dc segment and in this manner accomplishing various degrees of disintegration with Daubechies (db4) wavelet. The proposed technique is applied for the identification of epileptic case from given informational indexes as shown in Figure 8.1.

Data normalization is one of the key aspects during neural network training. The input variables with high value may tend to suppress the influence

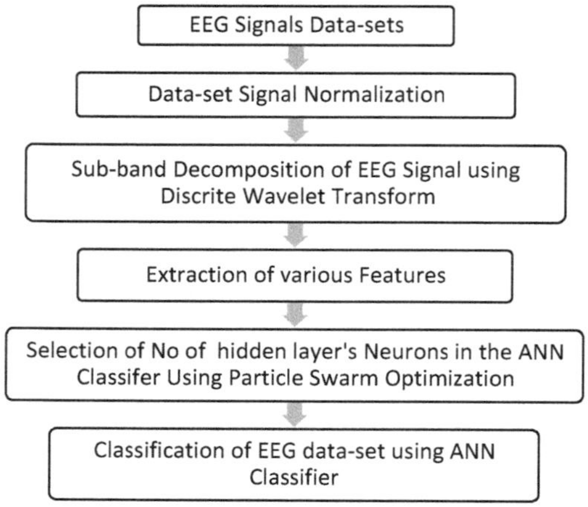

Figure 8.1 Steps in the classification of EEG dataset.

of smaller ones if processed without normalization. To rectify this problem, input data firstly are normalized and then neural networks are trained with input data. The variable value is normalized in the range between 0.9 and 0.1 accordingly. The normalized value x_n is calculated by using equation (8.1) [8]:

$$x_n = \frac{x - x_{\min}}{x_{\max} - x_{\min}} \times 0.8 + 0.1, \tag{8.1}$$

where x, $x_{\min}$, and $x_{\max}$ are real, minimum, and maximum values of input variable, respectively.

The next step is data segmentation. In this process, a window length of 256 discrete samples is selected from each dataset to form a single EEG segment. A total of 100-time series samples are present in each class, and each time series signal has 16 number of EEG segments, resulting in a total of 1,600 EEG segments prepared from each class dataset. Accordingly, from the available three class datasets, a total no of 4,800 segmented EEG signals have arranged.

For signal analysis using wavelet transform [9,10], appropriate selection of the mother wavelet and suitable decomposition level is of utmost significance. The wavelet decomposition is done by Daubechies wavelet (order 4, i.e., db4), because it has the best correlation with the EEG signal and its smoothing features are progressively reasonable to distinguish EEG signal's changes. The EEG dataset signals were applied under decomposition process till fourth level, and four details coefficients, i.e., D1–D4, and one approximation coefficient A4 are obtained. After determining these

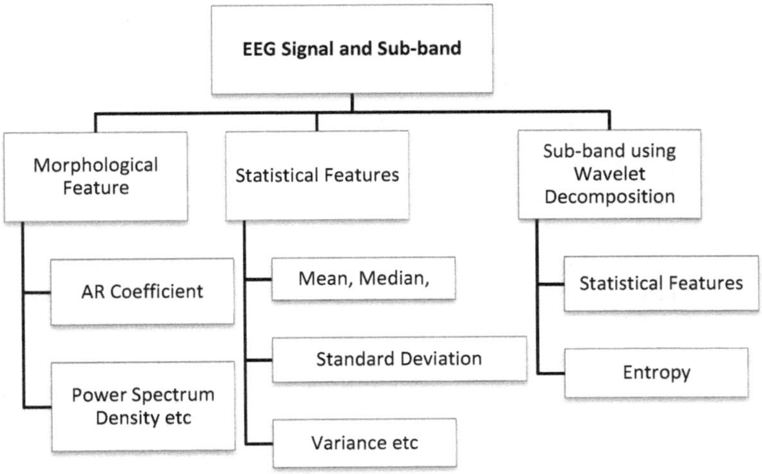

Figure 8.2 Various features extraction from EEG sub-band.

coefficients, the features extracted during the analysis are classified as shown in Figure 8.2. Various features have extracted from available dataset and arranged to form a feature vector having 87 features for the samples of 4,800 data segments. After feature extraction, we optimized the number of hidden layer neurons for ANN classifiers using the hybrid soft computing approach of ANN and PSO with or without TVAC and time-varying inertial weight (TVIW).

8.4 SELECTION OF HIDDEN LAYER NEURONS USING PARTICLE SWARM OPTIMIZATION

After the selection of hidden layer neurons with or without TVAC and TVIW in the PSO for ANN classifiers, classification accuracy results for various classes of epileptic seizures are compared for the following two cases: Case I: traditional PSO without implementing the concept of TVAC and TVIW, and Case II: PSO with implementing the concept of TVAC and TVIW, to determine the number of hidden layer neurons with fitness function as shown in equation (8.2).

$$\text{Fitness function} = \alpha \times \frac{\text{No of correct classified patterns}}{\text{Total patterns}} + \beta \times \frac{\text{No of nonselected features}}{\text{Total features}} \tag{8.2}$$

8.4.1 Case I: traditional PSO

The PSO is an effortless and influential tool for the optimization of solutions as particles into problem space. This approach is inspired by birds movement in a flock for food search and implemented as one of heuristic search method for exploring the possible solution in population. In PSO, particles known as swarms represented as possible outcomes of objective functions which accumulate information through constructed array by particles' positions, respectively, from other particles. These particles update positions by using their velocities. Both the velocity and position of particles are updated by using their own and neighbors' experience in an influential manner [11].

In the d-dimensional search space, position vectors of ith particle may be represented as $X_i = (x_{i1}, x_{i2}, \ldots x_{id})$ and velocity vectors of the ith particle may be represented as $V_i = (v_{i1}, v_{i2}, \ldots v_{id})$. Based on the fitness function value recorded, the best previous position of the particle is represented by $pbest_i = (p_{i1}, p_{i2}, \ldots p_{id})$. In a group of particles, if particle gth is the best, then it is represented as $p\text{best}_g = \text{gbest} = (p_{g1}, p_{g2}, \ldots P_{gd})$. Each particle's position is modified by using its current velocity and deviation from *pbest* and *gbest*.

In the next $(k+1)$th iteration to evaluate fitness, each particle's modified position and velocity are determined by using the following equations:

$$v_{id}^{k+1} = C\begin{bmatrix} wX\ v_{id}^{k} + c_1 X\ \text{rand}_1 X\ \left(p\text{best}_{id} - x_{id}\right) \\ + c_2 X\ \text{rand}_2 X\ \left(\text{gbest}_{gd} - x_{id}\right) \end{bmatrix} \tag{8.3}$$

$$x_{id}^{k+1} = x_{id} + v_{id}^{k+1}. \tag{8.4}$$

Here w is inertia weight parameter to controls particle's capabilities of global and local exploration, C is constriction factor, c_1 represents cognitive coefficients, c_2 represents social coefficients, and rand_1 and rand_2 are arbitrary numbers ranging between 0 and 1. During initial exploration, larger inertia weight factor is used, and as the search proceeds, its value is reduced gradually. TVIW [12] concept was implemented as follows:

$$w = \left(w_{\max} - w_{\min}\right) \times \frac{\left(\text{iter}_{\max} - \text{iter}_{\min}\right)}{\text{iter}_{\max}} + w_{\min}, \tag{8.5}$$

where $\text{iter}_{\max}$ is the iteration's maximum value. The constant c_1 pushes particles toward local best position, and, on other hand, c_2 pushes toward the global best position. The values of these parameters are selected in the range between 0 and 4. The constriction factor is also introduced to improve the convergence of PSO algorithm [13,14].

$$C = \frac{2}{|2 - \varphi - \sqrt{\varphi^2 - 4\,\varphi}\,|} \tag{8.6}$$

The value of the φ lie in the range $4.1 \le \varphi \le 4.2$. As the parameter φ increases, the constriction factor (C) decreases, and the resulting convergence becomes lower because the diversity of population is reduced.

8.4.2 Case II: PSO with time-varying acceleration coefficients (TVAC)

Implementing TVIW in PSO, we can find optimal solution significantly faster, but, at the search end, due to the lack of diversity, its ability of fine-tuning optimum solution is suffered. Researchers have observed that parameter's fine-tuning based on the problem is one of the important challenges in PSO to achieve optimum solution efficiently and accurately [15,16]. According to Kennedy and Eberhart [11] a relatively higher value of cognitive component compared to the social component causes individual to explore a wider search space. Conversely, a relatively higher value of the social component can lead particles to converge prematurely to a local optimum solution. Most of the studies have acceleration coefficients at 2 to make average equal to 1 for given stochastic factors in equation (8.3) to navigate by the particle during half the time of search only.

Optimization methods based on population adopt the policy to promote individual roaming during initial search through whole search space avoiding local optima clustering. To achieve optimum solution effectively, encourage convergence toward global best during latter stages. This can be done by varying c_1 and c_2 (acceleration coefficients) with time. As search proceeds, cognitive component reduced on one hand, and, on other hand, social component increased. In the beginning, the cognitive component value on upper side and social component value on lower side to move particle search space, instead moving premature toward population best and as reach towards latter stages makes-cognitive component value on lower side and social component value on higher side to make convergence of particle towards global optima.

The acceleration coefficients are determined as follows [17]:

$$c_1 = \left(c_{1f} - c_{1i}\right) \times \frac{\text{iter}}{\text{iter}_{\text{max}}} + c_{1i} \tag{8.7}$$

$$c_2 = \left(c_{2f} - c_{2i}\right) \times \frac{\text{iter}}{\text{iter}_{\text{max}}} + c_{2i}, \tag{8.8}$$

where c_{1i} and c_{1f} are initial and final values of cognitive acceleration factors, and c_{2i} and c_{2f} are initial and final social acceleration factors, respectively.

8.5 CLASSIFICATION USING NEURAL NETWORK

Artificial Neural Network using the hybrid soft computing approach is employed for epileptic classification [18]. Two cases were considered: ANN and PSO with or without TVAC (time-varying acceleration coefficient and time-varying inertial weight (TVIW). The optimized number of hidden layer neurons for both the cases are as follows:

Case I: Without implementing TVAC and TVIW

a. First hidden layer's neurons=40
b. Second hidden layer's neurons=24

ANN (back-propagation) classifier model for Case I is shown in Figure 8.3.

Case II: With implementing the concept of TVAC and TVIW.

a. First hidden layer's neurons=18
b. Second hidden layer's neurons=20

ANN (back-propagation) classifier model for Case II is shown in Figure 8.4.

8.6 RESULT AND DISCUSSIONS

The data used for analysis are EEG data subset for both healthy and epileptic subjects available online at http://www.meb.uni-bonn.de/epileptologie/science/physik/eegdata.html and the real-time recordings of EEG using

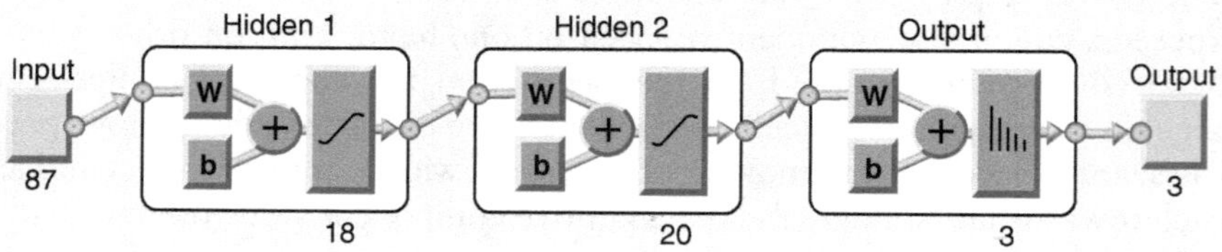

Figure 8.3 ANN model for Case I.

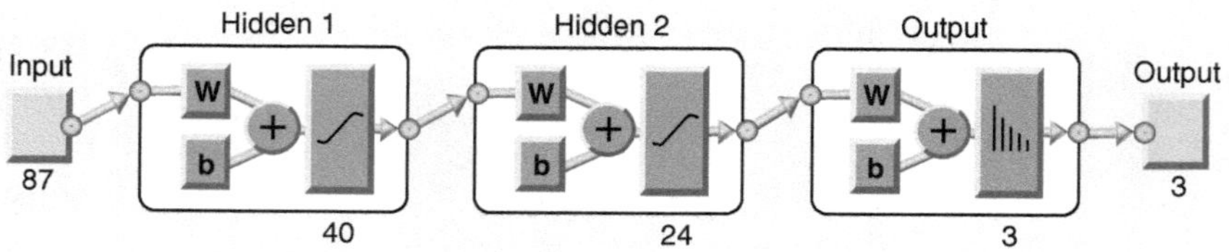

Figure 8.4 ANN model for Case II.

Table 8.1 The overall confusion matrix for Cases I and II

Dataset description	*Accuracy (%)*	
	Case I	*Case II*
Dataset-Z (normal person)	98.6	99.1
Dataset-F (epileptic patient under seizures-free conditions)	96.9	97.5
Dataset-S (epileptic patient during seizures)	97.7	98.6
Overall accuracy	97.8	98.4

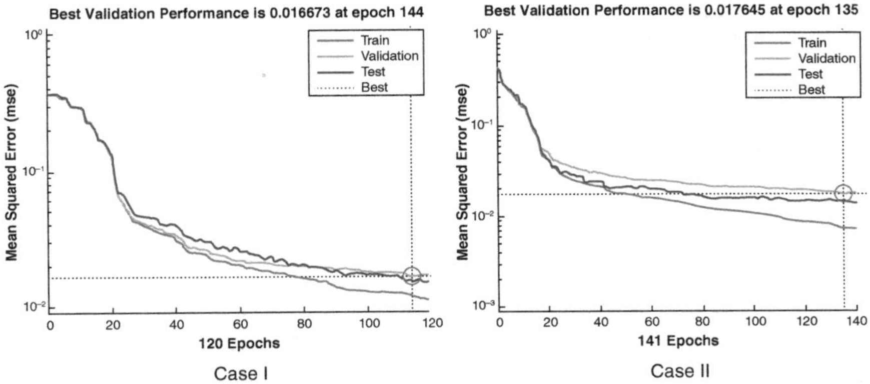

Figure 8.5 Training of artificial neural network: mean square error for Cases I and II.

wireless cap (using wireless sensors) in Biomedical Engineering Laboratory at MITS, Gwalior (MP). EEG's data-subset from three different groups, namely Dataset Z (normal), Dataset F (epileptic patients without seizure), and Dataset S (epileptic patients having seizure), are analyzed. These datasets have 100 single-channel recordings.

Each single-channel recording length is 26.3 s [19]. The datasets sample rate is 173.61 samples per second. In the single-channel recording, the total samples present are 4,097 (173.61×23.6). The neural network is trained by using these data as training and testing sets. Table 8.1 shows the overall confusion matrix results for the cases considered in this research work.

The MSE performance of the ANN for both cases are compared and are shown in Figures 8.5–8.7.

8.7 CONCLUSIONS

A MATLAB-based GUI analysis tool is designed and developed, and used for EEG dataset features extraction (statistical and morphological and segment dataset classification by using hybrid soft computing technique

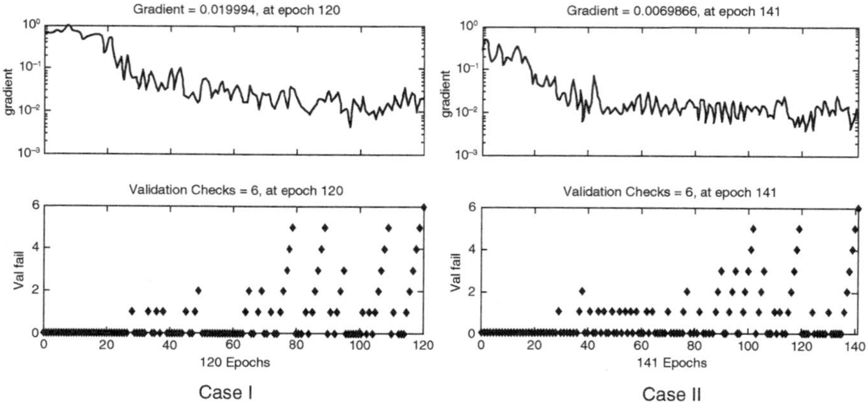

Figure 8.6 Training: validation check for Cases I and II.

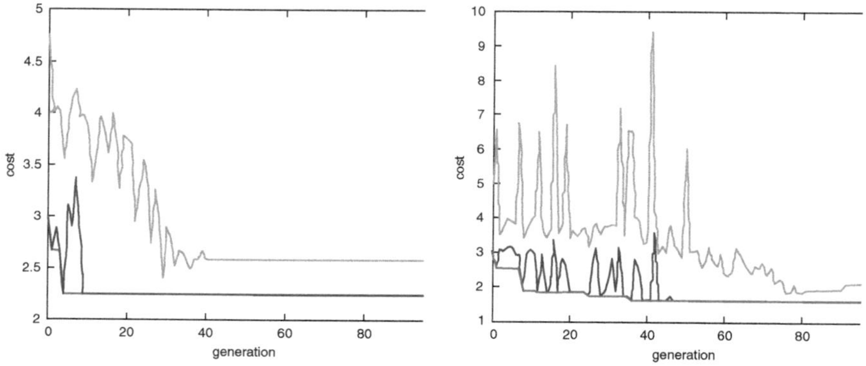

Figure 8.7 PSO results for Cases I and II.

of PSO and ANN). Epileptic classifications results are compared using PSO with and without implementing the concept of TVAC and TVIW for the selection of number of hidden layer neurons in the ANN classifier. Classification process implements ANN (back-propagation) using Levenberg–Marquardt training function for training. An overall efficiency of 98.4% is achieved for dataset classification with ANN classifier for Case II, i.e., with TVAC and TVIW classifier, as compared to Case I, i.e., without TVAC and TVIW, having an overall efficiency of 97.8%.

REFERENCES

1. Usakli Ali, B. (2009). Improvement of EEG signal acquisition: An electrical aspect for state of the art of front end. *Computational Intelligence and Neuroscience*. doi:10.1155/2010/630649
2. Alexandros, T., Markos, G. T., Dimitrios, G. T., Evaggelos, C. Karvounis, L.A., Spiros, K., & Margaret, T. (2011). Automated epileptic seizure detection methods: A review study. In: *Epilepsy: Histological, Electroencephalographic and Psychological Aspects*. doi:10.5772/31597
3. Übeyli, E. D. (2009). Statistics over features: EEG signals analysis. *Computers in Biology and Medicine*, 39(8), 733–741.
4. Iasemidis, L. D., Olson, J. C., Sackellares, & Savit, R. S. (1994). Time dependencies in the occurrences of epileptic seizures: A nonlinear approach. *Epilepsy Research*, 17(1), 81–94,
5. Adeli, H., Zhou, Z., & Dadmehr, N. (2003). Analysis of EEG records in an epileptic patient using wavelet transform. *Journal of Neuroscience Methods*, 123(1), 69–87.
6. Iasemidis, L. D. (2003). Epileptic seizure prediction and control. *IEEE Transactions on Biomedical Engineering*, 50(5), 549–558.
7. Guler, I., & Ubeyli, E.D. (2005). Adaptive neuro-fuzzy inference system for classification of EEG signals using wavelet coefficients. *Journal of Neuroscience Methods*, 148(2), 113–121. doi:10.1016/j.jneumeth.2005.04.013
8. Bandil, M.K., & Wadhwani, A. K. (2020). EEG based feature extraction & ANN classifier's neurons selection for epileptic classification using hybrid soft computing (ANN & GA). *Solid State Technology*, 63(5).
9. Ubeyli, E. D. (2008). Wavelet mixture of experts network structure for EEG signals classification. *Expert Systems with Applications*, 34(3), 1954–1962.
10. Polikar, R. (2006, November). *The Wavelet Tutorial*. Retrieved from https://cseweb.ucsd.edu//~baden/Doc/wavelets/polikar_wavelets.pdf
11. Kennedy, J., & Eberhart, R. (1995). Particle swarm optimization. In: *Proceedings of IEEE Conference on Neural Networks (ICNN'95)*, Perth, Australia, vol. 5, pp. 942–1948.
12. Shi, Y., & Eberhart, R. C. (1999). Empirical study of particle swarm optimization. *Proceedings of IEEE International Conference on Evolutionary Computation*, 3(1), 101–106.
13. Vlachogiannis, J. G., & Lee, K. Y. (2006). A comparative study on particle swarm optimization for optimal steady state performance of power systems. *IEEE Transactions on Power Systems*, 21(4), 1718–1727, 2006.
14. Eberhart, R. C., & Shi, Y. (2000). Comparing inertia weights and constriction factors in particle swarm optimization. In: *Proceedings of 2000 Congress Evolutionary Computation,* 16–19 July, La Jolla, CA, pp. 84–88.
15. Park, J. B., Lee, K. S., Shin, J. R., & Lee, K. Y. (2005). A particle swarm optimization for economic dispatch with non smooth cost functions. *IEEE Transactions on Power Systems*, 20(1), 34–42.

16. He, Q. H, Wu, J. Y, Wen, Saunders J. R., & Patton P. C. (2004). A particle swarm optimizer with passive congregation. *Biosystems*, 78,135–147.
17. Ratnaweera, A., Halgamuge, S. K., & Watson, H. C. (2004). Self Organizing hierarchical particle swarm optimizer with time-varying acceleration coefficients. *IEEE Transactions on Evolutionary Computation*, 8(3), 240–255.
18. Haykin, S. (1994). *Neural Networks: A Comprehensive Foundation*. New York: Macmillan.
19. Daubechies, I. (1990). The wavelet transform, time-frequency localization and signal analysis, *IEEE Transaction on Information Theory*, 36(5), 961–1005.

Chapter 9

Advancements in bilirubin detection

Harnessing smart sensor technologies for improved diagnosis

Kabita Kumari, Sharvan Kumar Pahuja, and Sanjeev Kumar

9.1 INTRODUCTION

Bilirubin, a pigment known for its distinctive golden color, plays a crucial function in the complex mechanisms of the human body. It is an important indicator of several physiological processes and is produced when the iron-containing molecule "heme," a component of hemoglobin in red blood cells, breaks down. The significance of jaundice extends beyond its visual appearance. It is a crucial indicator of liver health and the complex interaction between blood components. The identification of bilirubin in the blood stream by the German physician Rudolf Virchow in 1847 was a crucial breakthrough in comprehending the composition and the pathophysiology of blood. Virchow's meticulous blood examination identified the yellow pigment bilirubin as a notable constituent [1]. Bilirubin deposition in tissues occurs exclusively in the presence of elevated serum bilirubin levels, indicating either liver disease or, less frequently, a disorder in bilirubin metabolism. The extent of serum bilirubin elevation can be approximated through physical examination. Slight increases in serum bilirubin are best observed by examining the sclerae of the eyes. Sclerae, rich in elastin, have a specific affinity for bilirubin, and scleral icterus signifies a serum bilirubin level of at least 51 μmol/L (3 mg/dL). Detecting scleral icterus becomes more challenging in rooms illuminated with fluorescent lighting. Another reliable marker for elevated serum bilirubin is when the color of urine becomes dark, resulting from the excretion of conjugated bilirubin from urinary excretory system. Patients commonly characterize their urine as resembling tea or cola. The presence of bilirubin signifies an increase in the direct serum bilirubin fraction, indicating the existence of liver or biliary disease. The estimated bilirubin production in healthy adults is 3.5–4.0 mg/kg body weight per day. However, in newborn infants, it is twice as 8.5 ± 2.3 mg kg of body weight per day [2]. In human serum, there are two forms of bilirubin: unconjugated (Bu) and conjugated (Bc) [3]. When conjugated bilirubin combines with glucuronic acid, it forms a water-soluble compound, while unconjugated bilirubin binds with albumin, rendering it

DOI: 10.1201/9781003633884-9

water-soluble [4]. A minimal amount of unbound, free bilirubin not associated with albumin serves as a crucial indicator of bilirubin poisoning. Typically, the levels of direct bilirubin in human blood fall within the range of 0.06–0.3 mg/dL, and the total bilirubin concentration is approximately 1.23 mg/dL. Higher bilirubin levels may be a sign of jaundice, hepatitis, cirrhosis, blood infections, transfusion reactions, or hemolytic diseases in newborns that leads to cell destruction, Gilbert's syndrome, Crigler–Najjar syndrome, and other health problems that could be very dangerous. On the other hand, low bilirubin levels are connected to conditions like anemia and coronary artery diseases [5]. In neonates, untreated high concentrations of unconjugated free bilirubin can result in serious consequences such as brain damage, including hearing loss and the potentially fatal syndrome known as "Kernicterus" [6]. Checking bilirubin levels along with other liver enzyme tests can help to figure out if high bilirubin is caused by hemolysis or another liver disease that is not hemolysis. Pathological levels of bilirubin often coincide with the accumulation of its oxidized form, biliverdin, further emphasizing the diagnostic significance of bilirubin. Given this importance, the development of an affordable detection device for bilirubin becomes a matter of great interest as bilirubin is a valuable measure that can help with diagnosis and prognosis [7].

9.1.1 Chemical structure of bilirubin

The natural degradation of heme occurs through a sequence of intermediate stages, with the pivotal component being the orange-yellow bile pigment known as "bilirubin." This chromophore is accountable for the coloration observed in different types of jaundice. In young human erythrocytes have time about three months and their destruction causes a bilirubin production about 300 mg each day. But there were some early doubts arising (for cyclization between C18 vinyl group and the oxygen function at C19) for the gross chemical structure of bilirubin has been in reasonable disputes for many years [8]. The chemical structure of bilirubin was elucidated in 1937 by Fischer and Orth, revealing it as a tetrapyrrole with a close resemblance to hemoglobin [9]. Successful synthesis for bilirubin structure was reported in 1942 as shown in Figure 9.1a. Again in 1976, X-ray crystallography demonstrated that the structure took on a bis-lactam form, a finding further supported by 15N NMR spectroscopy studies in the 1980s [4,10], shown in Figure 9.1b. A simplification of the bis-lactum form of the bilirubin is presented again in Figure 9.1c. The bilirubin structure presented by Fischer and Plieninger in Figure 9.1 lacked details on stereochemistry, leading to inconsistent descriptions over the years, with variations between the correct structure 4Z,15Z and 4E,15E configurations. Subsequent studies have firmly established that bilirubin predominantly exists as the 4Z,15Z isomer, and this configuration may hold significance for process to entry into the brain and neurotoxicity caused by it [11].

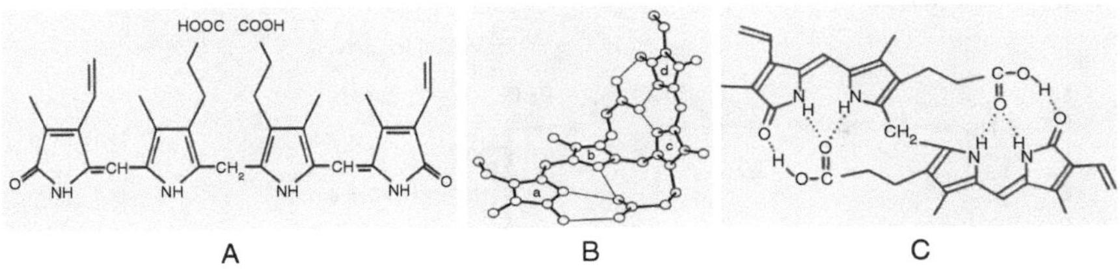

Figure 9.1 Structure of bilirubin. (a) The planar structure of bilirubin described by Fischer and Plieninger. (b) The bis-lactam structure of bilirubin by Bonnett, Davies, and Hursthouse, with hydrogen bonds indicated by thin lines. (c) A simplified version of structure b [11].

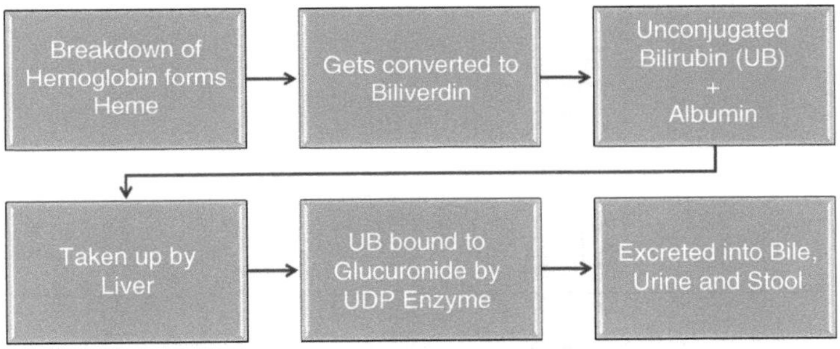

Figure 9.2 Block representation for metabolic pathway of bilirubin formation and excretion from body.

9.1.2 Metabolism of bilirubin

The journey of bilirubin metabolism commences with the natural breakdown of red blood cells, liberating heme into the bloodstream. Figure 9.2 delineates the metabolic pathway of bilirubin. Heme, predominantly sourced from aging red blood cells, contributes to about 20% of daily bilirubin production, while proteins like cytochrome P450 isoenzymes and myoglobin also play a role. Monocytic macrophages in the spleen and bone marrow, as well as hepatic Kupffer cells, synthesize heme [12]. Through hemo-oxygenase, heme is transformed into biliverdin, a linear molecule comprising four pyrrole rings, with the concomitant generation of iron and carbon monoxide [13]. Biliverdin reductase converts biliverdin into unconjugated bilirubin. To prevent isomerization and facilitate safe transportation throughout the body, bilirubin binds to albumin with a dissociation constant (K_d) of approximately 107–108 mol/L. Via sinusoids, albumin-bound bilirubin enters the liver. Inside liver cells, bilirubin binds to water-soluble proteins known as ligandins or Y proteins, inhibiting the excretion of absorbed bilirubin. In the smooth endoplasmic reticulum,

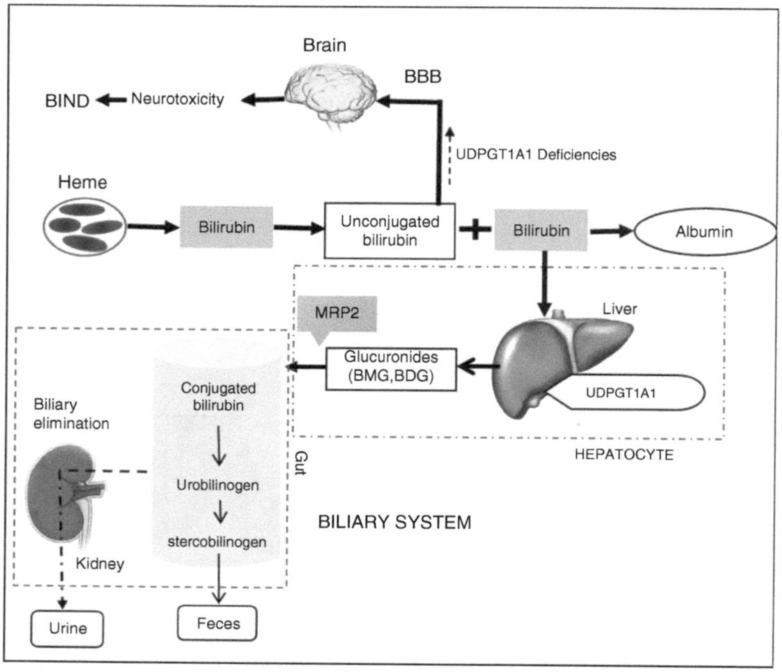

Figure 9.3 Schematic representation of metabolic process of bilirubin [17].

bilirubin undergoes conjugation with glucuronic acid by UDPGT-1A1, yielding bilirubin glucuronides [14]. Figure 9.3 illustrates the comprehensive phases of bilirubin production and excretion. Conjugated bilirubins return to the cytosol and can be excreted into bile or plasma through the canalicular membrane or sinusoidal membrane, respectively, facilitated by MRP2 (multidrug-related protein-2) [15]. The energy-intensive process of excreting conjugated bilirubin results in bile being significantly more concentrated than hepatocyte cytoplasm. Due to its water solubility, conjugated bilirubin is less prone to reabsorption by the gut, aided by microbial and intestinal mucosa β-glucuronidase. Unconjugated bilirubin may revert to its unconjugated form through hydrolysis, facilitated by β-glucuronidase. This unconjugated bilirubin can be reabsorbed by the intestinal mucosa, reentering the enterohepatic circulation. Roughly 25% of excreted bilirubin is reabsorbed via the bile duct [16]. Within the colon, anaerobic gut bacteria convert the remaining unconjugated bilirubin into colorless tetrapyrroles, such as stercobilinogen, urobilinogen, or mesobilinogen, which are collectively referred to as urobilinogens. Approximately 20% of the urobilinogen generated each day is absorbed back from the gut, contributing to enterohepatic recirculation. The majority of this urobilinogen is absorbed by the liver via the portal vein and then eliminated back into bile. Approximately 2%–5% of the substance reaches the bloodstream and is then excreted by the kidneys, making it detectable in urine.

9.1.3 Why clinical estimation of bilirubin is needed?

Clinical estimation of bilirubin is important for several reasons in the medical field, particularly in the assessment of liver and hematological disorders. Conditions that disrupt the normal metabolism of bilirubin can lead to elevated concentrations of this pigment in the blood vessels. When the concentration of bilirubin in the blood exceeds 1 mg/dL, it can result in the deposition of bilirubin in various tissues, especially those abundant in elastic fibers like the palate and conjunctiva of eyes [18]. A significant accumulation, typically surpassing 2.5 mg/dL, presents a yellowish skin and mucous membranes color, a condition commonly recognized as "jaundice." Jaundice is a clinical sign of hyperbilirubinemia. Elevated levels of bilirubin chromophore in the bloodstream can lead to its buildup in the brain as it crosses the blood–brain barrier. This condition, known as kernicterus, characterized by a yellow-stained nucleus, has the potential to cause brain damage. The impairment of brain function occurs through an unknown mechanism and may result in cerebral palsy and if left untreated, it can be fatal. An elevated level of bilirubin has affected adults as well as neonates. It mostly became vital in case of neonates due the improper development of neonate's liver. Neonatal jaundice has a notable impact on a significant portion of newborns, affecting around 60% of babies born at full term and 80% of those born prematurely [19,20]. Approximately 140 million newborns worldwide are affected with jaundice during the first two weeks of birth due to its high incidence [21]. Bhutani et al. [19], as illustrated in Figure 9.4a, were the first to present a mathematical modelling framework in pioneering the evaluation of the worldwide impact of severe jaundice was conducted. The model's estimation suggests that around 18%, i.e., approximately 24 million among 134 million instances of live births in the year 2010, were affected by jaundice of notable clinical significance.

Furthermore, approximately 481,000 neonates born between late-preterm and full-term experienced severe hyperbilirubinemia (total serum bilirubin [TSB] >25 mg/dL), leading to a total of 114,000 deaths and a survival rate exceeding 63,000 persons with lasting neurological issues ranging from moderate to chronic. Figure 9.4b represents the global estimation of impairment caused by kernicterus again in the year 2013, i.e., a total of 480,700 live births were found to be vulnerable to Rh illness within the cohort, 24% faced the risk of deaths for newborns (estimated at 114,100 with an uncertainty range of 59,700–172,000), 13% were susceptible to kernicterus (amounting to 75,400 with an uncertainty range of 43,200–95,800), and 11% were at risk of stillbirths [19]. It is essential to adhere to the guidelines set by the American Academy of Pediatrics (AAP) and

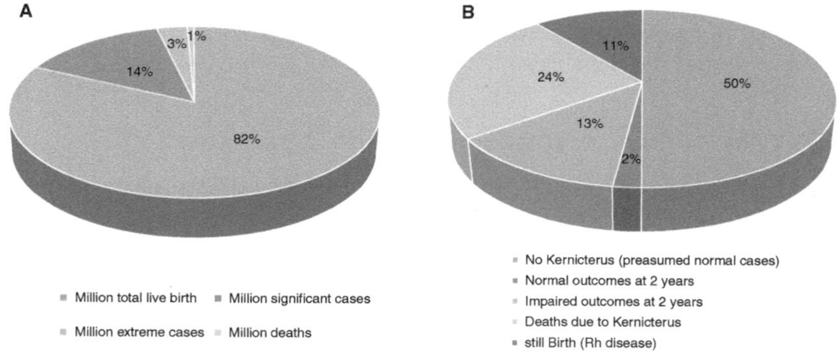

Figure 9.4 (a) Global occurrence of substantial jaundice in neonates [25] and (b) global consequences of persistent impairment resulting from kernicterus [26].

the UK's National Institute for Health and Care Excellence (NICE), which emphasize the importance of closely monitoring bilirubin levels in newborns and applying suitable treatments as needed [22–24]. Jaundice in newborn babies poses a noteworthy public health challenge in Asian countries like India, given the high birth rate and the insufficiently developed healthcare infrastructure. To address this issue, the Indian government has launched various initiatives like the Mission for Rural Health at the National Level (NRHM), aimed at training healthcare personnel and equipping primary health centers with phototherapy devices.

A scheme by government "Janani Shishu Suraksha Karyakram" complements these efforts by providing complimentary care and treatment for newborns with jaundice has proven effective in diminishing the prevalence of neonatal Jaundice. However, continuous research and development are essential. The focus should be on developing more accurate and economical diagnostic instruments to facilitate early detection and intervention. In summary, clinical estimation of bilirubin is a valuable tool in the diagnosis and monitoring of various medical conditions, especially those related to liver function and hematology. It furnishes vital data for healthcare practitioners to make well-informed judgments regarding patient care and treatment.

9.2 DIFFERENT APPROACHES FOR THE DETECTION BILIRUBIN

There are several analytical methods presently accessible for quantifying bilirubin and its byproducts, i.e., metabolites in serum, urine, and fecal samples. Various invasive and noninvasive methods are available for detecting jaundice. The measurement of TSB is a prevalent invasive technique for assessing hyperbilirubinemia levels. This method relies on spectrophotometry and involves obtaining a small blood sample using a syringe and needle [27]. This method is considered as gold standard method for bilirubin estimation. A general invasive process being followed in the clinical laboratories is shown in Figure 9.5. Nonetheless, the assessment of TSB is limited to skilled medical professionals in well-equipped facilities. Additionally, given its invasive nature, this procedure induces stress for both infants and their parents. Since initial description of bilirubin was in the late 19th century, the predominant techniques for measuring serum bilirubin rely on the bilirubin's reaction with diazotized sulfanilic acid, commonly referred to as the diazo reaction [2].

Numerous bilirubin tests utilized in clinical chemistry laboratories employ diazotization and oxidase methodologies [28]. Under highly acidic conditions with a pH ranging between 1.7 and 2.0, bilirubin and diazo ions undergo a colorimetric reaction, forming a red azo compound through an automated process. While the diazo test for bilirubin is cost-effective and automated, it is vulnerable to hemolysis and can be affected by lipids and albumin interference, potentially resulting in pH fluctuations and unreliable bilirubin measurements [29,30]. The existing bilirubin diazo detection method is only suitable for qualitative analysis because of its inability to meet the stringent accuracy benchmarks needed for precise quantitative assessment [31,32]. Bilirubin oxidizing enzyme is categorized as a multi-copper oxidase within the oxidoreductase class. Utilizing metal ions, it facilitates the oxidation of bilirubin into biliverdin, making it a commonly employed reagent in bilirubin testing [33]. Another technique involves "chemiluminescence," where bilirubin undergoes redox reactions in aqueous solutions [34]. This process involves the reaction between

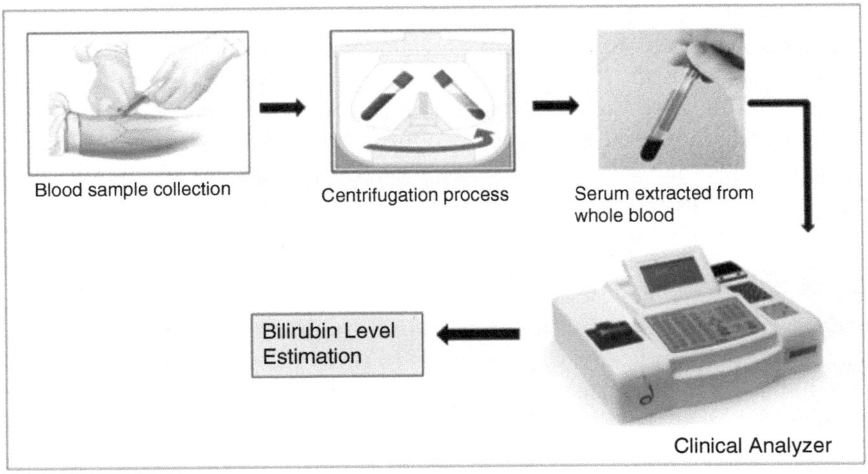

Figure 9.5 Laboratory estimation of bilirubin value measurement (invasive method).

N-bromosuccinimide and sodium hypochlorite as oxidants, resulting in the emission of intense radiation. A batch chemiluminometer was utilized to analyze these chemiluminogenic reactions. Mechanized analysis of the reactivity between N-bromosuccinimide, hypochlorite, and bilirubin was performed using flow injection and continuous flow analyzers. This method demonstrated sensitivity, with a broader linearity range (0.2–20 µg/mL) and a lower detection limit (50 ng/mL). However these processes are invasive in nature, and this approach to measure serum bilirubin concentrations (TSB) entails a painful process of blood sampling [35], a procedure that carries various potential long-term repercussions such as the rare occurrence of infection and osteomyelitis [36,37]. Healthcare workers are also exposed to the risk of needle stick injury with blood sampling. So, to overcome these problems, an alternative way, i.e., noninvasive methods, can be employed or further advancement can be made.

9.3 NONINVASIVE APPROACHES FOR BILIRUBIN DETECTION

Various developments have been made and some are ongoing for the noninvasive or contactless assessment of bilirubin. In this section, a discussion on the existing array of noninvasive approaches and methodologies employed as point-of-care screening and testing devices for the quantification of bilirubin as well as diseases (e.g., jaundice) caused due to bilirubin levels is presented.

9.3.1 Visual inspection

Visual assessment is a common method for estimating bilirubin levels, involving the observation of skin and mucous membrane coloration. The yellow discoloration, known as jaundice, indicates elevated bilirubin in the blood and serves as a straightforward, noninvasive means of identifying hyperbilirubinemia in newborns. One assessment method, the Kramer scale, evaluates jaundice severity based on skin appearance, following the cephalocaudal progression principle. This principle notes that jaundice typically starts on the face in newborns and spreads to the chest, abdomen, and extremities, as TSB levels rise [14]. Researchers Purcell and Beeby looked into whether changes in skin temperature and blood flow across certain body parts (forehead, ribs, lower belly, mid-thigh, and sole) were linked to babies getting jaundice from their heads to their feet. The study discovered that babies often have more blood flow to their heads and nearby areas in their first few days of life. This causes temperature to rise and jaundice to build up in these areas [38]. However, accurately assessing jaundice severity in newborns receiving outpatient care presents challenges. Some newborns may have significant hyperbilirubinemia without obvious clinical jaundice manifestations. Relying solely on the head-to-toe progression may delay severe jaundice identification and management, risking neurological complications like kernicterus. Additionally, this approach overlooks ethnic and racial differences, potentially exacerbating severity among certain patient populations [39–41].

9.3.2 Transcutaneous bilirubinometry

Laboratory-based testing for TSB is the preferred method, although it might be difficult in resource-limited areas owing to high costs and a lack of trained staff. Noninvasive transcutaneous bilirubin (TcB) devices can now be used for screening instead of looking at someone with the naked eye in some cases. These gadgets use visible light to check the skin and figure out serum bilirubin levels, which gives an approximate estimate of blood bilirubin levels. The forehead and sternum are often used places for testing. Although TcB devices come in various designs, they share a common feature of analyzing skin reflectance spectra, which involves examining the light absorbed and reflected by the skin and subcutaneous tissue when exposed to different wavelengths. The composition of cutaneous chromophores, such as a dark-brown pigment, i.e., melanin, collagen (a fibrous protein found in connective tissue), oxygenated and reduced hemoglobin, and notably bilirubin, influences the range of light wavelengths, i.e., spectra of the reflected light [42,43]. Table 9.1 is a compilation of current studies focusing on six distinct types of noninvasive TcB devices. The table includes information such as the title of the research, the year of publication, the instrument name, the technique or method used, the place or site of measurement, and the conclusion of the study.

Kuo et al. [44] pioneered a noninvasive method to detect bilirubin levels by spectrally analyzing vivid visuals utilizing a compact LED light source. These visuals were captured with a smartphone and analyzed for the intensity of red (R), green (G), and blue (B) hues. The study aimed to explore the direct correlation between variations in direct bilirubin levels and the spectral characteristics of the test paper image. Notably, the green channel exhibited the highest coefficient of determination "*R*" at 0.9313 and the lowest limit of detection at 0.56 mg/dL, falling within the direct bilirubin concentration span of 0.1–2.0 mg/dL. The "AJONEO" device is a novel development in the field of TcB meters. It is designed to take measurements from the newborn nail bed. An observational study was done in Kolkata, India, over a period of 15 months. The study included 2,092 newborn babies with gestational period between 28 and 40 weeks. The study identified the effective population size as 1,968 newborns. The study found a strong positive linear connection between AJONEO and TSB bilirubin concentrations, with no significant variations seen between right and left nail bed measures [37]. Another noninvasive device came into picture was Bilicheck. The BiliChek™ was created by SpectRx Inc. (Norcross, GA) and is today

Table 9.1 Recent studies in the development of noninvasive measurement method

S. No.	*Title*	*Year*	*Sample size*	*Name of device*	*Measurement technique*	*Site of measurement*	*Conclusion*
1	Detecting direct bilirubin noninvasively by analyzing color pictures spectrally using a mini-LED light source [44]	2022	12 direct bilirubin tests	NA	Spectrum analysis of colored images	NA	Quantifies direct above 1.86 mg/dL bilirubin rapidly and noninvasively
2	Impact of assessment location on TcB measurement [45]	2013	58 full term babies	Bilicheck	Transcutaneous	Forehead and sternum	Accurate readings at sternum than forehead
3	Computer vision for noninvasive early jaundice diagnosis [46]	2023	225 images	NA	Clustering, filtration, and novel metrics based on color space	Sclera	Accuracy of 88.57% is obtained
4	Evaluating the Bilicare TcB's accuracy as a screening tool for healthy-term and late-preterm neonates before discharge [47]	2018	2014 paired samples	Bilicare	Transcutaneous	NA	TSB readings are overestimated at 12 mg/dl and underestimated at higher TSB levels
5	Assessment of novel transcutaneous bilirubinometers in neonates [48]	2022	141 full-term neonates	JAISY v/s JM-105	Transcutaneous	Forehead and sternum	Precise in measuring low to moderate levels of bilirubin
6	Developing a mobile phone camera-based technology for low-resource environments [49]	2022	37 images	Mobile images	Transcutaneous	NA	Exhibits promise but needs enhancement and further research

widely utilized and promoted by Philips Electronics (Amsterdam, The Netherlands). The BiliChek™ equipment requires the use of a one-time-use tip that needs to be connected to the probe before doing a TcB measurement. This increases the overall cost of operating the instrument. A cohort study was conducted on 58 full-term babies who did not have hemolytic illness. Bilicheck was used to perform transcutaneous measurements on the forehead and sternum. The correlation and concordance between both techniques and plasma bilirubin were computed. A robust positive connection was seen between the measurements of serum bilirubin at the forehead and sternum with correlation values (r) of 0.704 with p¡0.01 and 0.653 with p¡0.01, respectively [45]. Sreedha et al.'s [46] proposed method for diagnosing jaundice (elevated bilirubin levels) is noninvasive. The proposed method can effectively detect and precisely estimate the extent of illness severity by analyzing the level of yellowness in the sclera. The weighted average of normal (healthy) and jaundiced (unhealthy) eyes at different levels of the disease is calculated by analyzing several eye pictures. The efficacy of the suggested method is assessed by conducting tests on a dataset consisting of 35 eye pictures, and the results are then validated. Among the 35 test photos, the suggested approach successfully assessed the severity of jaundice with accuracy in 31 photographs. In the remaining four photos, the technique was unsuccessful due to an inadequately created mask using the semantic segmentation model. The proposed approach has a testing data accuracy of 88.57%. Peerathat et al. [47] presented a study the effectiveness and the accuracy of Bilicare transcutaneous meter in late-preterm and term neonates was determined. A cross-sectional study was conducted where inclusion criteria consisted of healthy neonates who were born at or after 35 weeks of gestation. The blood sample for TSB measurement was collected simultaneously with the normal metabolic test done within the 48–72-hour period prior to discharge. The association between TSB levels varied between 2.53 and 16.69 mg/dL, with an average (standard deviation) of 9.79 (2.83) mg/dL. The TcB levels ranged from 3.10 to 16.60 mg/dL, with a mean of 10.49 (2.47) mg/dL.

An evaluation of performance of a novel transcutaneous bilirubinometer called JAISY was presented by Norman et al. [48]. A total of 930 bilirubin measures were conducted on 141 newborn children who were delivered near-term or at term and were between 1 and 6 days old. Using an earlier developed and validated device JM-105, the obtained results of these measurements were then compared. The measured bilirubin levels ranged from 0 to 18.8 mg/dL. There was a strong association between bilirubin measurements obtained from the two devices on the forehead (correlation coefficient (r) = 0.94) and the chest (r value = 0.94). Harrison-Smith et al. [49] presented a mobile phone-based TcB technique and validated it with a 37-patient multiethnic pilot trial. An adapter for filtered red-green-blue (RGB) mobile phone camera reflectance measurements was designed using Monte Carlo simulations of neonatal skin reflectance. This showed a link between TSB and mobile phone TcB estimated bilirubin with R_2=0.42 and Bland–Altman limits of agreement of +6.4 to −7.0 mg/dL. Various research studies have been conducted and are ongoing in the development of noninvasive technique as point-of-care devices for bilirubin estimation. The adoption of TcB has grown prevalent in childbirth centers and outpatient departments due to its prompt, easy, and nonintrusive and painless characteristics. However, it is crucial to admit that TcB determination serves as a diagnostic screening tool, and values nearing therapeutic thresholds (e.g., initiation of phototherapy) should be verified through TSB measurements. While having various applications and advantages, Tcb devices also possess some drawbacks in terms of skin pigmentation and race, ambient light may affect the output, affect of phototherapy on TcB reading, etc. [50,51]. Further research needs to be carried out to establish the long-term benefits and applications of TcB screening. Nonetheless, documented positive short-term outcomes provide a foundation for ongoing investigation.

9.4 CONCLUSION

In summary, jaundice is a common issue in neonatal care, often appearing in the first week of life and leading to frequent early hospital readmissions. Current standard practices for assessing bilirubin levels involve invasive TSB measurements, considered the gold standard, and noninvasive TcB assessments using specialized meters. While TSB is direct and invasive, TcB aims to estimate jaundice through optical signals on blanched skin, reducing the need for invasive procedures and minimizing distress for newborns and parents. Noninvasive approaches should complement assessment rather than replace treatment. Future studies on TcB should give primary importance to enhancing the accuracy of devices, while considering factors such as pigmentation of the skin, prenatal age, and postnatal age. Consider utilizing innovative methods such as multispectral imaging and fluorescence spectroscopy to estimate bilirubin levels in infants. With advancements in electronic devices, there is potential for a more reliable and safe method for measuring serum bilirubin in newborns. The growing trend of health sensing through smartphones opens up the possibility of a noninvasive device using smartphones as a cost-effective and accurate alternative for managing neonatal jaundice in clinical settings.

REFERENCES

1. Guirguis N, Bertrand A, Rose CF, Matoori S. 175 years of bilirubin testing: ready for point-of-care? *Advanced Healthcare Materials* 2023;12:1–8. https://doi.org/10.1002/adhm.202203380.

2. Maisels MJ, Pathak A, Nelson NM, Nathan DG, Smith CA. Endogenous production of carbon monoxide in normal and erythroblastotic newborn infants. *Journal of Clinical Investigation* 1971;50:1–8. https://doi.org/10.1172/JCI106463.
3. Fevery J. Bilirubin in clinical practice: a review. *Liver International* 2008;28:592–605. https://doi.org/10.1111/j.1478-3231.2008.01716.x.
4. Taurino I, Van Hoof V, De Micheli G, Carrara S. Superior sensing performance of multi-walled carbon nanotube-based electrodes to detect unconjugated bilirubin. *Thin Solid Films* 2013;548:546–550. https://doi.org/10.1016/j.tsf.2013.09.015.
5. Sherlock S, Dooley J, Iber F. *Diseases of the Liver and Biliary System*. Wiley, New York, 2002.
6. Ahlfors CE, Parker AE. Evaluation of a model for brain bilirubin uptake in jaundiced newborns. *Pediatric Research* 2005;58:1175–1179. https://doi.org/10.1203/01.pdr.0000185248.43044.cd.
7. AuYoung B, Gutha Ravichandran A, Patel D, Dave N, Shah A, Wronko-Stevens B et al. A low-cost paper-based device for the colorimetric quantification of bilirubin in serum using smartphone technology. *Frontiers in Chemistry* 2022;10:1–11. https://doi.org/10.3389/fchem.2022.869086.
8. Bonnett R, Davies J, Hursthouse MB. Structure of bilirubin. *Nature* 1976;262:326–328. https://doi.org/10.1038/262326a0.
9. Fischer H, Plieninger H. Synthese des biliverdins (uteroverdins) und bilirubins, der biliverdine XIII α und III α, sowie der Vinylneoxanthosäure. *Hoppe-Seyler's Zeitschrift Für Physiologische Chemie* 1942;274:231–260. https://doi.org/10.1515/bchm2.1942.274.1-6.231.
10. Hansen PE, Jakobsen HJ. A natural abundance 15 N NMR investigation of bilirubin IX-α. *Organic Magnetic Resonance* 1984;22:668–670. https://doi.org/10.1002/mrc.1270221015.
11. Hansen TWR, Wong RJ, Stevenson DK. Molecular physiology and pathophysiology of bilirubin handling by the blood, liver, intestine, and brain in the newborn. *Physiological Reviews* 2020;100:1291–1346. https://doi.org/10.1152/physrev.00004.2019.
12. Jacobsen J. Binding of bilirubin to human serum albumin: determination of the dissociation constants. *FEBS Letters* 1969;5:112–114. https://doi.org/10.1016/0014-5793(69)80307-8.
13. Erlinger S, Arias IM, Dhumeaux D. Inherited disorders of bilirubin transport and conjugation: new insights into molecular mechanisms and consequences. *Gastroenterology* 2014;146:1625–1638. https://doi.org/10.1053/j.gastro.2014.03.047.
14. Rowland A, Miners JO, Mackenzie PI. The UDP-glucuronosyltransferases: Their role in drug metabolism and detoxification. *The International Journal of Biochemistry & Cell Biology* 2013;45:1121–1132. https://doi.org/10.1016/j.biocel.2013.02.019.
15. Kumari K, Pahuja SK, Kumar S. An evolution of bilirubin physiology and analysis. *Current Signal Transduction Therapy* 2023;18. https://doi.org/10.2174/1574362418666230216152920.
16. Guerra Ruiz AR, Crespo J, López Martínez RM, Iruzubieta P, Casals Mercadal G, Lalana Garcés M et al. Measurement and clinical usefulness of bilirubin in liver disease. *Advances in Laboratory Medicine/Avances En Medicina de Laboratorio* 2021;2:352–361. https://doi.org/10.1515/almed-2021-0047.

17. Kumari K, Pahuja SK, Kumar S. An evolution of bilirubin physiology and analysis. *Current Signal Transduction Therapy* 2023;18:213734. https://doi.org/10.2174/1574362418666230216152920.
18. Ichihara K, Ozarda Y, Barth JH, Klee G, Qiu L, Erasmus R et al. A global multicenter study on reference values: 1. Assessment of methods for derivation and comparison of reference intervals. *Clinica Chimica Acta* 2017;467:70–82. https://doi.org/10.1016/j.cca.2016.09.016.
19. Bhutani VK, Stark AR, Lazzeroni LC, Poland R, Gourley GR, Kazmierczak S et al. Predischarge screening for severe neonatal hyperbilirubinemia identifies infants who need phototherapy. *The Journal of Pediatrics* 2013;162:477–482.e1. https://doi.org/10.1016/j.jpeds.2012.08.022.
20. Amos RC, Jacob H, Leith W. Jaundice in newborn babies under 28 days: NICE guideline 2016 (CG98). *Archives of Disease In Childhood - Education and Practice* 2017;102(4):207–209. doi:10.1136/archdischild-2016-311556.
21. UNICEF. *The State of the World's Children 2017.* Children in a Digital World, UNICEF, New York, 2017.
22. NICE, *Neonatal Jaundice.* National Institute for Health and Care Excellence (NICE), National Collaborating Centre for Women's and Children's Health, London, 2010. https://doi.org/10.7748/ncyp.28.6.11.s15.
23. American Academy of Pediatrics Subcommittee on Hyperbilirubinemia, Management of hyperbilirubinemia in the newborn infant 35 or more weeks of gestation. *Pediatrics* 2004;114:297–316.
24. World Health Organization, *United Nations Children's Fund WBG. Postnatal Care for Mothers and Newborns: Highlights from the World Health Organization 2013 Guidelines.* Geneva: World Health Organization, 2015, pp. 1–8.
25. Abiha U, Banerjee DS, Mandal S. Demystifying non-invasive approaches for screening jaundice in low resource settings: a review. *Frontiers in Pediatrics* 2023;11. https://doi.org/10.3389/fped.2023.1292678.
26. Mori R, Slusher TM, Fahmy N, Paul VK, Du L, Okolo AA et al. Neonatal hyperbilirubinemia and Rhesus disease of the newborn: incidence and impairment estimates for 2010 at regional and global levels. *Pediatric Research* 2013;74:86–100. https://doi.org/10.1038/pr.2013.208.
27. Khanam FTZ, Al-Naji A, Perera AG, Wang D, Chahl J. Non-invasive and non-contact automatic jaundice detection of infants based on random forest. *Computer Methods in Biomechanics and Biomedical Engineering: Imaging & Visualization* 2023;11:2516–2529. https://doi.org/10.1080/21681163.2023.2244601.
28. Ellairaja S, Shenbagavalli K, Ponmariappan S, Vasantha VS. A green and facile approach for synthesizing imine to develop optical biosensor for wide range detection of bilirubin in human biofluids. *Biosensors and Bioelectronics* 2017;91:82–88. https://doi.org/10.1016/j.bios.2016.12.026.
29. Singha SS, Mondal S, Bhattacharya TS, Das L, Sen K, Satpati B et al. Au nanoparticles functionalized 3D-MoS2 nanoflower: An efficient SERS matrix for biomolecule sensing. *Biosensors and Bioelectronics* 2018;119:10–17. https://doi.org/10.1016/j.bios.2018.07.061.
30. Dhungana N, Morris C, Krasowski MD. Operational impact of using a vanadate oxidase method for direct bilirubin measurements at an academic medical center clinical laboratory. *Practical Laboratory Medicine* 2017;8:77–85. https://doi.org/10.1016/j.plabm.2017.05.004.

31. Devgun MS, Richardson C. Direct bilirubin in clinical practice—Interpretation and haemolysis interference guidance reassessed. *Clinical Biochemistry* 2016;49:1351–1353. https://doi.org/10.1016/j.clinbiochem.2016.09.005.
32. Kazmierczak SC, Robertson F, Catrou PG, Briley KP, Kreamer B, Gourley G. Direct spectrophotometric method for measurement of bilirubin in newborns: comparison with HPLC and an automated diazo method. *Clinical Chemistry* 2002;48:1096–1097.
33. Rawal R, Kharangarh PR, Dawra S, Bhardwaj P. Synthesis, characterization and immobilization of bilirubin oxidase nanoparticles (BOxNPs) with enhanced activity: Application for serum bilirubin determination in jaundice patients. *Enzyme and Microbial Technology* 2021;143:109716. https://doi.org/10.1016/j.enzmictec.2020.109716.
34. Palilis LP, Calokerinos AC, Grekas N. Chemiluminescence arising from the oxidation of bilirubin in aqueous media. *Analytica Chimica Acta* 1996;333:267–275. https://doi.org/10.1016/0003-2670(96)00279-6.
35. Yap SH, Mohammad I, Ryan CA. Avoiding painful blood sampling in neonates by transcutaneous bilirubinometry. *Irish Journal of Medical Science* 2002;171:188–190. https://doi.org/10.1007/BF03170276.
36. Yüksel S, Yüksel G, Öncel S, Divanli E. Osteomyelitis of the calcaneus in the newborn: an ongoing complication of Guthrie test. *European Journal of Pediatrics* 2007;166:503–504. https://doi.org/10.1007/s00431-006-0268-z.
37. Halder A, Adhikari A, Ghosh R, Singh S, Banerjee A, Ghosh N et al. Large scale validation of a new non-invasive and non-contact bilirubinometer in neonates with risk factors. *Scientific Reports* 2020;10:11149. https://doi.org/10.1038/s41598-020-67981-9.
38. Purcell N, Beeby PJ. The influence of skin temperature and skin perfusion on the cephalocaudal progression of jaundice in newborns. *Journal of Paediatrics and Child Health* 2009;45:582–586. https://doi.org/10.1111/j.1440-1754.2009.01569.x.
39. Devi S, Dash M, Chitra F. Detection of neonatal jaundice among the newborn using Kramer's criteria. *Epidemiology: Open Access* 2018;8. https://doi.org/10.4172/2161-1165.1000355.
40. Sampurna MTA, Mapindra MP, Mahindra MP, Ratnasari KA, Rani SAD, Handayani KD et al. Kramer score, an evidence of its use in accordance with Indonesian hyperbilirubinemia published guideline. *International Journal of Environmental Research and Public Health* 2021;18:6173. https://doi.org/10.3390/ijerph18116173.
41. van der Geest BAM, de Mol MJS, Barendse ISA, de Graaf JP, Bertens LCM, Poley MJ et al. Assessment, management, and incidence of neonatal jaundice in healthy neonates cared for in primary care: A prospective cohort study. *Scientific Reports* 2022;12:14385. https://doi.org/10.1038/s41598-022-17933-2.
42. Engle WD, Jackson GL, Engle NG. Transcutaneous bilirubinometry. *Seminars in Perinatology* 2014;38:438–451. https://doi.org/10.1053/j.semperi.2014.08.007.
43. Jacques SL, Saidi IS, Ladner A, Oelberg D. Developing an optical fiber reflectance spectrometer to monitor bilirubinemia in neonates. In: Jacques SL, editor. *Laser-Tissue Interaction* VIII, vol. 2975, 1997, pp. 115–124. https://doi.org/10.1117/12.275465.

44. Kuo H-C, Ye Z-T, Tseng SF, Tsou SX, Huang SW, Tsai C-W. Noninvasive direct bilirubin detection by spectral analysis of color images using a Mini-LED light source. *Discover Nano* 2023;18:16. https://doi.org/10.1186/s11671-023-03794-9.
45. Conceição CM, Dornaus MFPS, Portella MA, Deutsch AD, Rebello CM. Influence of assessment site in measuring transcutaneous bilirubin. *Einstein (São Paulo)* 2014;12:11–15. https://doi.org/10.1590/S1679-45082014AO2711.
46. Sreedha B, Nair PR, Maity R. Non-invasive early diagnosis of jaundice with computer vision. Procedia Computer Science 2022;218:1321–1334. https://doi.org/10.1016/j.procs.2023.01.111.
47. Chokemungmeepisarn P, Tantiprabha W, Kosarat S, Manopunya S. Accuracy of the Bilicare™ transcutaneous bilirubinometer as the predischarge screening tool for significant hyperbilirubinemia in healthy term and late preterm neonates. *The Journal of Maternal-Fetal & Neonatal Medicine* 2020;33:57–61. https://doi.org/10.1080/14767058.2018.1484098.
48. Norman M, Aytug H, Celebi HB. Evaluation of a new transcutaneous bilirubinometer in newborn infants. *Scientific Reports* 2022;12:5835. https://doi.org/10.1038/s41598-022-09788-4.
49. Harrison-Smith B, Dumont AP, Arefin MS, Sun Y, Lawal N, Dobson D et al. Development of a mobile phone camera-based transcutaneous bilirubinometer for low-resource settings. *Biomedical Optics Express* 2022;13:2797. https://doi.org/10.1364/BOE.449625.
50. Yamauchi Y, Yamanouchi I. Factors affecting transcutaneous bilirubin measurement: effect of daylight. *Pediatrics International* 1991;33:658–662. https://doi.org/10.1111/j.1442-200X.1991.tb01882.x.
51. Kumar D, Kumar D. A prospective comparison of serum and transcutaneous bilirubin in Indian neonates. *Journal of Pediatric Intensive Care* 2022;11:100–104. https://doi.org/10.1055/s-0040-1721067.

Chapter 10

Development of internet of things-based system for home automation using soft computing

Praveen Kumar Malik and Arshi Naim

10.1 INTRODUCTION

To design home appliance control with the help of internet or the mobile phone is the revolution. Data are made easily accessible to the user by a webpage's user-friendly graphical user interface [1]. Applications like smart homes, smartphones, smart cities, smart watches, and smart shops have become indispensable in today's technologically evolved society. There is a critical need to save energy and save costs in whatever manner feasible due to the growing population [2]. A platform for connecting, sensing, and remotely controlling numerous equipment across a network infrastructure is offered by internet of things (IoT). In the context of this study, home automation refers to the use of smartphones and computers to operate the electrical, electronic, and mechanical appliances in many types of buildings. An administrator can connect devices to a cloud server to enable multiple users with a large number of sensor and control nodes to control those devices [3].

The plan involves using a website or smartphone to remotely control electronic gadgets and home appliances over the internet. Additionally, voice recognition technology is used to support those who are physically or visually unable [4]. The general idea is to construct automated homes using IoT-based sensing and monitoring technologies. The NodeMCU board, which can function as a mini web server and interface with different hardware modules, is used in the prototype. The system provides switching capabilities for controlling lights, fans, and other home appliances through the relay system. The system can be controlled remotely through the internet by using an Android OS smartphone [5]. The history and application of the IoT are examined in this overview for house. According to a study [6], smart houses have advantages for managing energy and water use, comfort and convenience, and even cost savings for those who are financially vulnerable. Home automation's main objective is to make it easier to regulate domestic chores by automating them and enhancing comfort. Raspberry Pi,

DOI: 10.1201/9781003633884-10

relays, Atmel Board, and mobile phones are just a few of the instruments utilized in home automation. The chapter examines previous academics' use of the IoT in house control application research [7].

10.2 REVIEW OF LITERATURE

Smart homes reduce the amount of human involvement in monitoring and managing house settings and appliances. The chapter outlines a method for creating smart home applications by fusing cloud computing, web services, and the IoT applications. The strategy focuses on using the Arduino platform to incorporate intelligence into sensors and actuators, connecting nodes using Zigbee RF, and connects to the cloud with live updates [8,9]. Technology for automation is developing. In every way, life is becoming simpler and simpler. House with internet-enabled systems with wireless technology with the help of IoT is a system that uses a Microcontroller (MCU) and handles them from anywhere from the globe. In this work, the microcontroller is a Raspberry Pi. The goal of the work to develop an automated work that can assist the devices in making wiser and can reduce power consumption by switching to Pulse Width Modulation (PWM) and switch actuating devices. Additionally, a security system using Infra Red (IR) sensors will be developed to increase the level of security in the home [10].

The goal is to create a home automation system that can operate linked electrical appliances and carry out all the fundamental tasks of a virtual assistant, including telling the time, date, and temperature. There is no need to input anything because the entire system is voice-activated. In addition to using voice instructions, the system will also use a few sensors to automate a few appliances [11].

The IoT allows us to monitoring and control of specific activities using sensors and actuators on the machines that are a part of a system or infrastructure. The model for remote device control via hand gestures is presented in this study. The gestures are identified using template matching algorithms to operate remote items via the internet of items in accordance. The Local Area Lan (LAN) or Wi-Fi module is used to connect the distant devices and microcontrollers used in the architecture to the internet [12]. Home automation systems allow us to inspect, manage, and regulate all of our household services and allowances manually, without the assistance of any other machinery or human labor. This chapter discusses several methods for manually controlling all of the programs in our house while utilizing a remote control. Every human being, but especially those who are physically challenged, benefits greatly from these practices in society [13]. House automation utilizes an internet-connected house to link electronic gadgets together and enables remote management of such items using an

Android smartphone and web client (server). Applications with an Arduino microcontroller and Ethernet shield put on an Android smartphone. In this study, Arduino serves as the easiest server for data control or retrieval via an Ethernet network. Data requests and arrangements can be made using a web client or an Android application with UDP process [14].

In order for customers to be able to operate the system from any distance, the initial generation of home automation equipment must be connected to the internet. IoT has become necessary for automation because of this. Lighting, temperature, atmosphere, entertainment, and many more items may all be controlled by a home automation system. In addition to security control and alarm systems, automation also incorporates home security [15,16]. Automation theory, technological development, and smart phones with Android operating systems have led to refined and intelligent lifestyles. A smarter home network results from the implementation of automatic switching with the IoT. This study designs and implements a smart home automation IoT-based prototype [17–19].

10.3 BLOCK DIAGRAM

As per the Figure 10.1, the work's focal point is the ESP8266 NuttyFi Wi-Fi Board. It is a development board built on the ESP8266 that offers Wi-Fi connection. The AC lights are managed by the two 12V 2-channel relays. The relays are connected to the AC lamps via AC cables. An external power source powers the ESP8266 NuttyFi Wi-Fi Board, the relays, and the AC

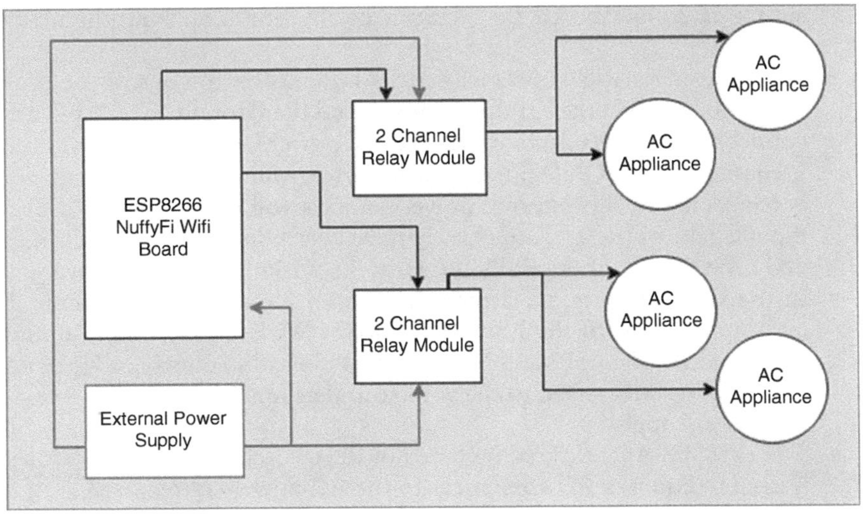

Figure 10.1 Block diagram of the NuttyFi-based node.

lighting. Jumper wires are used to connect the work's numerous components. The ESP8266 NuttyFi Wi-Fi Board is intended to control the two 12V 2-channel relays. The relays are connected to the AC lighting. The AC lighting will be turned on by the ESP8266 NuttyFi Wi-Fi Board by signaling the relays. The AC lighting will stay on till the ESP8266 NuttyFi Wi-Fi Board is turned off.

A web browser or a mobile app may be used to remotely operate the ESP8266 NuttyFi Wi-Fi Board. The ESP8266 NuttyFi Wi-Fi Board has to be set up with a Wi-Fi network and an IP address in order to accomplish this. Once set up, the ESP8266 NuttyFi Wi-Fi Board is accessible from any device linked to the same Wi-Fi network. The user may use a web browser and enter the IP address of the ESP8266 NuttyFi Wi-Fi Board to remotely control the AC lamps. After that, a web page with an on/off switch for the AC lamps will be displayed to the user. The AC lamps may also be controlled by the user via a mobile app. To operate ESP8266-based devices, there are several smartphone applications available. This work may be used to operate a number of AC appliances, including heaters, fans, and lights. The work may also be used to build a remote-controllable smart home system.

10.4 HARDWARE DEVELOPMENT

Figure 10.2 depicts the connection of the hardware using NuttyFi:

- The internet is used by the system's main controller, the ESP8266 NuttyFi Wi-Fi Board, to connect it to the IoT app. The external power source is connected to the Vcc pin of the ESP8266 NuttyFi Wi-Fi Board.
- We are using a 2-channel relay in two quantities to control the four AC appliances via the IoT app. Voltage and the Ground (GND) pin are connected to the Ground of the external power source. As a result, the 2-channel relay's GND pin is connected to ground, and its voltage pin is connected to the external power source's voltage. The input 1 and input 2 pins of the 2-channel relay have been connected to the GPIO8 and GPIO7 pins of the ESP8266 NuttyFi Wi-Fi Board, respectively.
- In the same manner, we attach the second 2-channel relay's input 1 and input 2 pins to the ESP8266 NuttyFi Wi-Fi Board's GPIO6 and GPIO5 pins. The ESP8266 NuttyFi Wi-Fi board's voltage and ground pins are similarly wired to the voltage and ground pins of the everlasting power supply.
- The external AC terminal included in the two-channel relay modules is used to connect AC appliances to the AC power source.
- With the aid of this invention, we may regulate any AC equipment in accordance with consumer demands.

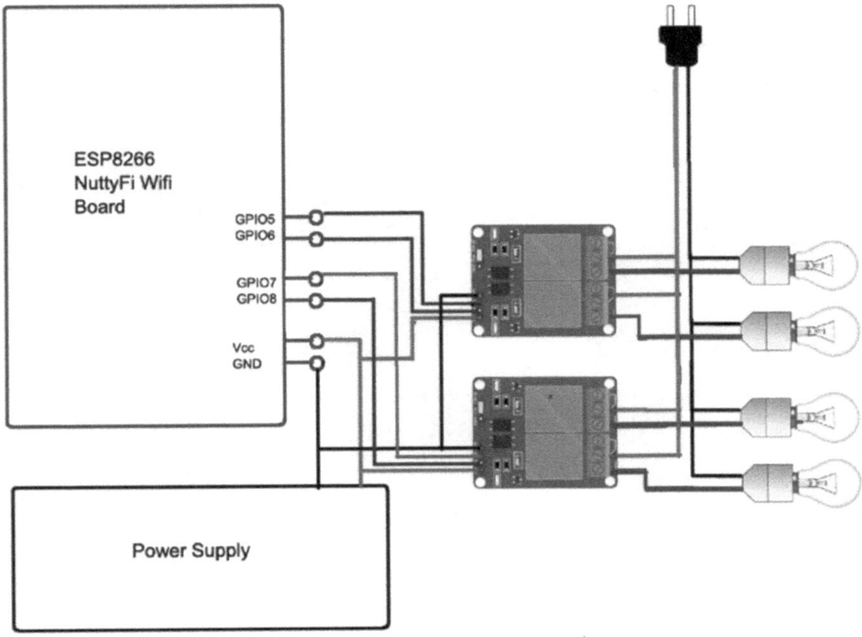

Figure 10.2 Connection diagram of NuttyFi-based mote.

10.4.1 Implementation of the system

The ESP8266 NuttyFi Wi-Fi Board, two 12V 2-channel relays, AC lamps, and other required components will be used to create a home automation system for this work. The initial steps of the algorithm involve setting up the ESP8266 board and connecting it to the Wi-Fi network at home. The General Purpose Input Output (GPIO) pins and the relays are controlled by libraries and dependencies that have been installed and set up (Figure 10.3). The GPIO pins are configured to control the relays in charge of turning on and off the AC bulbs. Remote control and monitoring are made possible via the ESP8266 Board's communication with a cloud platform or a local server. Users can communicate with the system by using a user interface. In order to determine the required action, the algorithm continuously listens for commands and decodes and interprets them.

By delivering the proper signals to the GPIO pins, relays can be activated or deactivated in response to user commands. Mechanisms for error management and feedback are in place to handle uncommon cases and offer status updates. Sensor integration is an option that permits automatic responses to outside events.

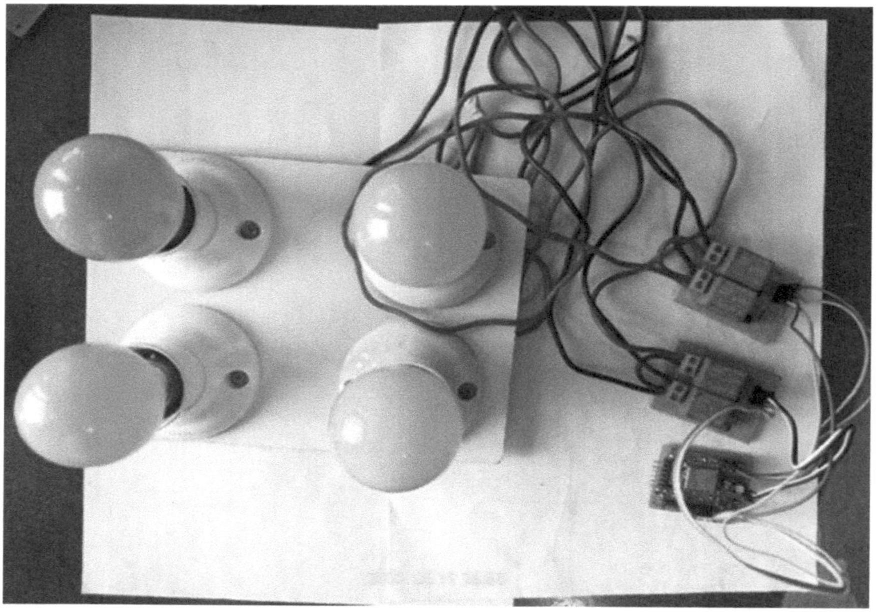

Figure 10.3 NuttyFi-based mote.

Security elements like authentication and encryption guard the system. Extensive testing ensures dependability, responsiveness, and functionality. Through the dashboard of the Blynk IoT mobile app, users may monitor the results of real-time data mapping and control (Figure 10.4).

10.5 CONCLUSION

An ESP8266 NuttyFi Wi-Fi Board may be used to remotely operate AC equipment. Two 12V 2-channel relays can be set to be controlled by the ESP8266 NuttyFi Wi-Fi Board. Any AC equipment, including heaters, fans, and lights, can be connected to the relays. A web browser or a mobile app can be used to remotely control the ESP8266 NuttyFi Wi-Fi Board. With the help of the proposed method, a smart home control system might be built. This work offers several advantages. By turning off AC gadgets while they are not in use, energy can be saved. By enabling customers to control AC devices remotely, even when they are not at home, it can increase security. By enabling users to manage AC devices with a single button press, it may make life more convenient. In general, this work is an excellent approach to remotely operate AC equipment. It is simple to set up and use, and it has several advantages.

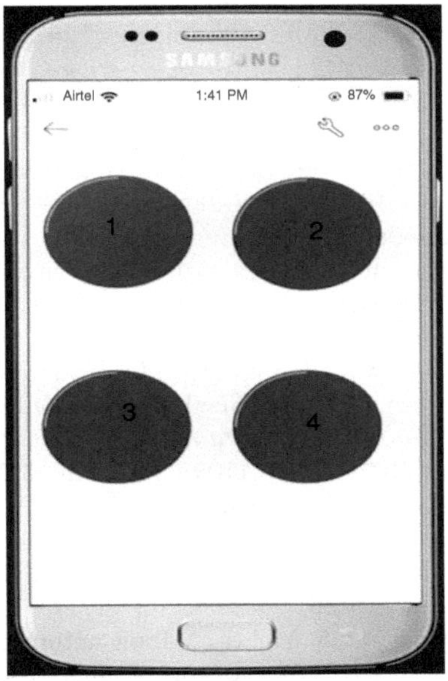

Figure 10.4 Blynk2.0 IoT web dashboard.

REFERENCES

1. M. D'souza, N. Wilfred, R. Pereira, T. Rayen, and A. M. Telgote, "Home automation using internet of things," in *2017 International Conference on Energy, Communication, Data Analytics and Soft Computing (ICECDS)*, 1 August 2017, doi:10.1109/ICECDS.2017.8389498.
2. S. Majumder, S. Ray, C. Ghosh, and S. Datta, "Usage of internet of things in home automation systems: a review," in *International Conference on Modeling, Simulations and Optimizations*, 2020, doi:10.1007/978-981-15-9829-6_5.
3. S. Dey, A. Roy, and S. Das, "Home automation using internet of thing," in *2016 IEEE 7th Annual Ubiquitous Computing, Electronics & Mobile Communication Conference (UEMCON)*, 1 October 2016, doi:10.1109/UEMCON.2016.7777826.
4. M. Kushwah and R. Rani, "Home automation and security system using internet of things," in *Advances in Wireless Technologies and Telecommunication*, 2019, doi:10.4018/978-1-5225-7335-7.CH012.
5. K. L. Raju, V. Chandrani, S. S. Begum, and M. P. Devi, "Home automation and security system with node MCU using internet of things," in *2019 International Conference on Vision Towards Emerging Trends in Communication and Networking (ViTECoN)*, 1 March 2019, doi:10.1109/ViTECoN.2019.8899540.

6. K. Moser, J. Harder, and S. G. M. Koo, "Internet of things in home automation and energy efficient smart home technologies," in *2014 IEEE International Conference on Systems, Man, and Cybernetics (SMC)*, 4 December 2014, doi:10.1109/SMC.2014.6974087.
7. A. K. N. Susan, N. S. Ayare, N. S. Prema, and S. Chandini, "Internet of things in home automation: a review," in *Advances in Intelligent Systems and Computing*, 2019, doi:10.1007/978-981-13-8618-3_79.
8. M. Soliman, T. Abiodun, T. Hamouda, J. Zhou, and C.-H. Lung, "Smart home: integrating internet of things with web services and cloud computing," in *2013 IEEE 5th International Conference on Cloud Computing Technology and Science*, 2 December 2013, doi:10.1109/CloudCom.2013.155.
9. Y. Arora, H. V. Pant, and B. Banita, "Home automation system with the use of internet of things and artificial intelligence," in *2019 International Conference on Innovative Sustainable Computational Technologies (CISCT)*, 1 October 2019, doi:10.1109/CISCT46613.2019.9008167.
10. U. Painuli, G. Mishra, A. Aggarwal, A. Irfan, and V. Sharma, "A review of home automation technology installation with the help of raspberry Pi and internet of things," *Journal of Mobile Technologies, Knowledge, and Society*, vol. 7, no. 2, pp. 22–27, 2020.
11. A. M. Jain, P. Tanwar, and S. Mehra, "Home automation system using internet of things (IOT)," in *2019 International Conference on Machine Learning, Big Data, Cloud and Parallel Computing (COMITCon)*, 1 February 2019, doi:10.1109/COMITCon.2019.8862201.
12. S. Khatri, G. Sharma, P. Johri, and S. Mohan, "Smart gesture control for home automation using internet of things," in *2018 International Conference on Computer, Communication, and Signal Processing (ICCCSP)*, 2018, doi: 10.1007/978-981-10-7245-1_62.
13. S. Supriya, R. Charanya, and S. Madhumitha, "A review on home automation system using IOT," in *2020 International Conference on Emerging Trends in Information Technology and Engineering (ic-ETITE)*, 1 February 2020, doi:10.1109/ic-ETITE47903.2020.363.
14. N. Dengen, R. M. Yusuf, and Pohny, "Internet of things: design and manufacture of home automation system with web server based Arduino and Android," *International Journal of Recent Technology and Engineering*, vol. 8, no. 2, pp. 3447–3453, 2019, doi:10.35940/ijrte.b1311.0982s1119.
15. P. Mercy, "Implementation and control of IoT based home robotixation using Arduino," *International Journal for Research in Applied Science and Engineering Technology*, vol. 7, no. 2, pp. 514–517, 2019, doi:10.22214/IJRASET.2019.2054.
16. K. K. Rout, S. Mallick, and S. Mishra, "Design and implementation of an internet of things based prototype for smart home automation system," in: *2018 International Conference on Recent Innovations in Electrical, Electronics & Communication Engineering (ICRIEECE)*, 1 July 2018, doi:10.1109/ICRIEECE44171.2018.9008410.
17. N. Shaik and P. K. Malik, "A comprehensive survey 5G wireless communication systems: open issues, research challenges, channel estimation, multi carrier modulation and 5G applications," *Multimedia Tools and Applications*, 2021. doi:10.1007/s11042-021-11128-z

18. N. Shaik and P. K. Malik. "A retrospection of channel estimation techniques for 5g wireless communications: opportunities and challenges," *International Journal of Advanced Science and Technology*, vol. 29, no. 5, pp. 8469–8479, 2020.
19. S. V. Akram, P. K. Malik, R. Singh, A. Gehlot, A. Juyal, K. Z. Ghafoor, and S. Shrestha, *Implementation of Digitalized Technologies for Fashion Industry 4.0: Opportunities and Challenges*, Scientific Programming, Hindawi, 2022, doi:10.1155/2022/7523246

Chapter 11

Integration of smart sensors in DFIG-based wind energy conversion systems

Neeraj Kumar Mishra, Jayprakash Giri, and Manoj Kumar Shukla

11.1 INTRODUCTION

The wind energy conversion (WEC) system is becoming a demanding area for renewable energy generation. The worldwide energy demand increases, as well as the demand for smart sensors, for doubly fed induction generator (DFIG) technology. Due to the utilization of this DFIG technology, new challenges like reliability, cost of operation, faults, and vibration occur in the machine [1]. During the physical operation of the machine, number of problems like open circuits and short circuit faults of stator and rotor windings, and front and back bearing fault occur in DFIG technology [2]. The integration of smart sensors in DFIG-based WEC systems detect the fault at an early stage and simultaneously warn the system for appropriate action. The smart sensors can measure different parameters of electric machines like speed, angle, vibration, and temperature. This monitoring is used for the early discovery of faults or abnormalities. In addition, smart sensors can sense problems in real time, for example, disturbances in the grid or component failures [3].

The smart sensors can provide a much larger amount of data that can be used for advanced analytics and machine learning systems. A data-driven approach helps in schedule optimization, predicting components life span, and improving the overall system performance. Sensors can be programmed to set off alarms and shut down operations when the environment becomes unsafe, thereby minimizing the opportunities for accidents. In the current scenario, the demand for smart sensors is raising because they take advantage of conventional and emerging AI-based algorithms [4].

The real-time simulator used for hybrid renewable energy systems contains normal and smart sensors [5]. These sensors perform four main functions:

- **Measurement:** It involves detecting physical signals and converting them into electrical signals. This helps monitor and measure things like temperature, traffic flow, and industrial processes.

DOI: 10.1201/9781003633884-11

- **Configuration:** It is important because it allows the smart sensor to detect its position and correct installation errors.
- **Verification:** It involves continuous monitoring of sensor behavior using internal circuits or equipment within the sensor.
- **Communication:** It enables the sensor to interact with the main microcontroller or microprocessor.

The sensors are divided into two parts, namely normal sensor and smart sensor. Normal sensors are devices that detect and measure physical properties such as temperature, pressure, light, motion, or humidity, and convert these into electrical signals, while smart sensors are advanced versions of normal sensors that can perform data processing, decision-making, and communication functions, with integrated processing capabilities. Figure 11.1 illustrates the block diagram representation of normal sensors and smart sensors. The important example of normal sensors are thermocouples, strain gauges, and LDRs.

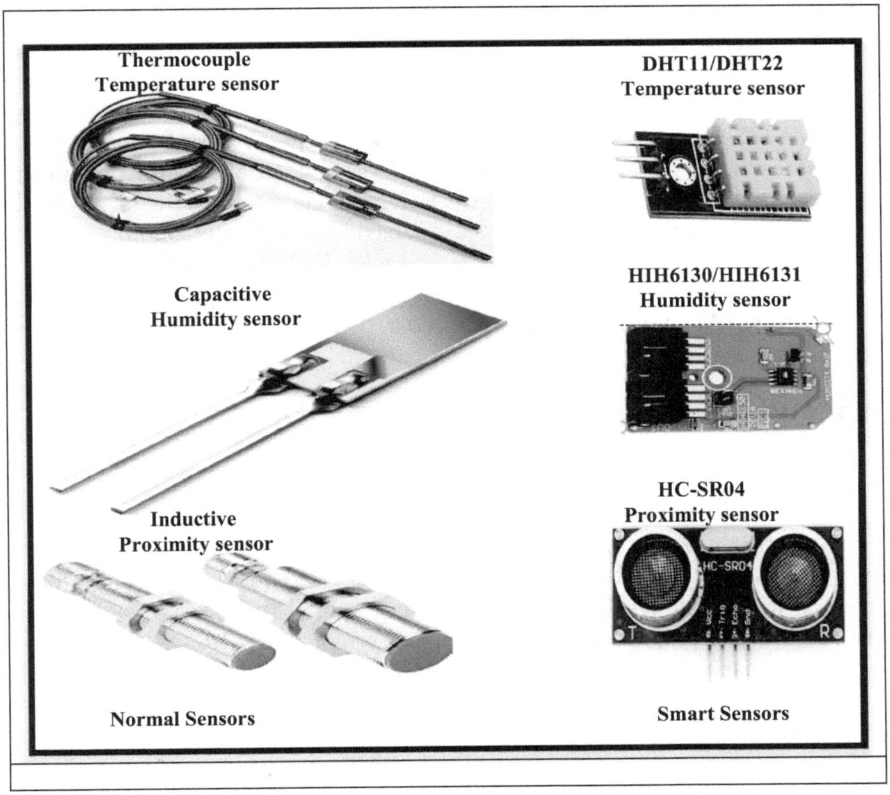

Figure 11.1 Block diagram representation of normal sensors and smart sensors.

Table 11.1 Comparison of normal sensors and smart sensors

Criteria	*Normal sensor*	*Smart sensor*
Purpose	Detects physical and chemical changes in the environment	Used for computing and processing tasks
Digital motion processor (DMP)	Accelerometer, Gyroscope, Magnetometer	Includes DMP or digital motion processor
Components	Sensor element, packaging, connections, and signal processing hardware	Amplifiers, transducers, analog filters, excitation control, compensation sensors
Types of sensors	Pressure, position, temperature, vibration, force, humidity, fluid properties	Electric current, level, humidity, pressure, proximity, temperature, heat, flow, etc.
Output usability	Requires conversion into a usable format	Output is ready to use
Cost	Inexpensive due to fewer components	More expensive due to additional capabilities
Use case	Provides sensor input for devices requiring complete control	Offers native processing capabilities

- The important examples of smart sensors are smart thermostats, MEMS-based accelerometers, and smart cameras with built-in image recognition capabilities. The comparison of normal sensors and smart sensors is illustrated in Table 11.1. Normal sensor detects physical and chemical changes, suitable for basic sensor input. Smart sensor, used for computing and processing tasks, includes additional components like digital motion processor, ready-to-use output, and provides enhanced capabilities at a higher cost.

11.2 APPLICATIONS OF DIFFERENT TYPES OF SENSORS USED FOR DFIG

Integration of smart sensors in DFIG-based WEC systems requires rigorous condition monitoring for the proper functioning of the system, and the knowledge and application of different sensors are required. Some common sensors utilized in DFIG-based wind energy conversion systems (WECS) are current sensors, voltage sensors, speed sensors, position sensors, temperature sensors, vibration sensors, pressure sensors, flux sensors, optical sensors, power sensors, etc. Figure 11.2 illustrates the block diagram representation of different types of sensors.

I. **Current Sensors:** DFIG is a doubly excited device, the measurement of current in both windings, stator, and rotor is required. The accurate current measurement is important for the protection of faults

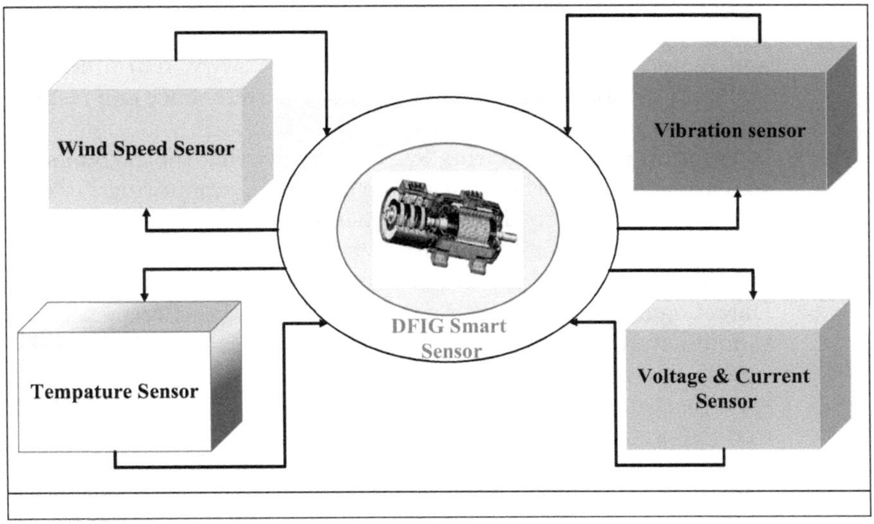

Figure 11.2 In this chapter, integration of smart sensors for WEC systems are presented.

from overcurrent current conditions. In addition, the current sensor also ensures the efficient operation and controlling the power output for WECS [6]. Some important types of current sensors are Hall effect sensors, current transformers, and Rogowski coils [7]. Smart current sensors hold a significant place in the list of important elements for keeping a watch over the electric currents in DFIGs. The sensors help in ensuring the right functioning of the generator and can provide data that could be useful for the diagnosis a generator's electrical problems.

Here are some features to consider when choosing smart current sensors for DFIG applications:

- **Measurement Range:** Choose current sensors that are capable of handling the currents that can be expected in the stator and rotor windings of the DFIG. Make sure that the sensors are capable of accurately measuring not only low but also high current values.
- **Accuracy and Precision:** Pick up sensors with good accuracy and precision so as to get accurate current measurements. The correct up-to-date data are fundamental to the proper control and safekeeping of the generator.
- **Response Time:** Fast response times, being one of the most important criteria for catching the dynamic changes of present, are essential. The sensors with fast response provide real-time data that lead to the rapid detection of unusual circumstances.

- **Wireless Connectivity:** Think about sensors with built-in wireless technology, such as Bluetooth or Wi-Fi, for the quick setup and remote control. Through Wi-Fi connectivity, real-time current data can be obtained and preventive maintenance can be performed online.
- **Integration with Monitoring Systems:** Verify that the current sensors are able to interface with wide condition monitoring systems or supervisory control and data acquisition (SCADA) systems. This integration enables a centralized monitoring of current data and allows us to address the asset management comprehensively.
- **Data Logging and Trend Analysis:** Besides, the current sensors should have the ability to log data to record current patterns over time. Some sensors are shipped with built-in or compatible software that allows you to analyze the historical data and get insights into the predictive maintenance.
- **Self-Diagnostics:** Make sure that you buy sensors with built-in self-diagnostic or check-ups that can monitor their performance. Self-diagnostics help with accurate readings and also actuate alarms if the sensor does not function properly.
- **Power Efficiency:** It is wise to select sensors with low power consumption to prolong battery life. Due to long-lasting batteries, the batteries' maintenance and replacement frequency are reduced, especially in places that are hard to reach or situated far away.
- **Certifications and Standards:** Prefer sensors that follow the industry standards and certifications, which are important factors in determining their reliability and suitability in critical applications like DFIGs.
- **Durability and Environmental Resistance:** Due to the fact that DFIGs operate under severe conditions, it is recommended to select current sensors that are resistant, durable, and do not break down due to various environmental factors such as moisture, dust, and temperature variations.
- **Overcurrent Protection:** As for some modern current sensors, they may have overcurrent protection features as well and so help prevent the generator's damage if the current conditions become unusual.
- **Alarm and Notification Systems:** Intelligent sensors of the present age must be able to produce alarms or notifications in a real-time mode whenever the current thresholds are exceeded. This is possible as the problem can be detected and fixed instantaneously to avoid damage or downtime.

By considering these characteristics, operators can select smart current sensors. Figure 11.3 illustrates the inductive current sensor, and Table 11.2 represents the purpose, functionality, and benefits of smart current sensors.

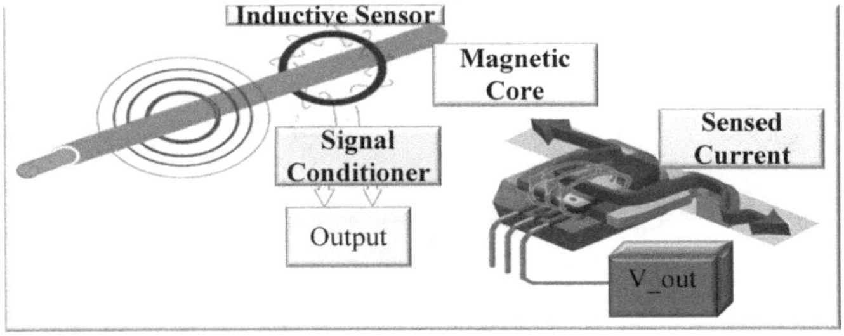

Figure 11.3 Smart inductive current sensors for DFIG.

Table 11.2 Purpose, functionality, and benefits of smart current sensors

Purpose	To monitor electrical current in the stator and rotor windings
Functionality	These sensors provide real-time data on current flow to detect abnormal conditions or faults
Benefits	These sensors enable precise control of the generator and help prevent electrical issues

II. **Voltage Sensors:** Voltage sensors in the DFIG-based WEGS scheme monitor the stator-side and rotor-side voltage levels. The output of sensing devices takes decisive action for controlling the active and reactive power of DFIG technology. In stator frequency regulation of DFIG requires two stator-side voltage sensors. Some important voltage sensors are capacitive voltage sensor, optical voltage sensors, potential transformers, Hall effect voltage sensors, and resistive voltage dividers. Nowadays, some sensor-less schemes have also been introduced [8–10].

Figure 11.4 illustrates the configuration of smart voltage sensors for DFIG. When selecting smart voltage sensors for DFIG applications, consider the following features:

- **Measurement Range:** Select voltage sensors that will be able to cover the voltage range of both stator and rotor windings of the DFIG. Provide the sensors with the ability to accurately measure the low as well as high voltages.
- **Accuracy and Precision:** Choose the sensors with high accuracy and precision to get the voltage readings that are accurate and reliable. Correct voltage data are necessary for those functions to be adequately performed and protected.
- **Response Time:** Instant response times play a major role, as they help the control system to track minor movements in voltage. Quick response sensors give actual data, which are necessary for timely abnormal voltage condition detection.

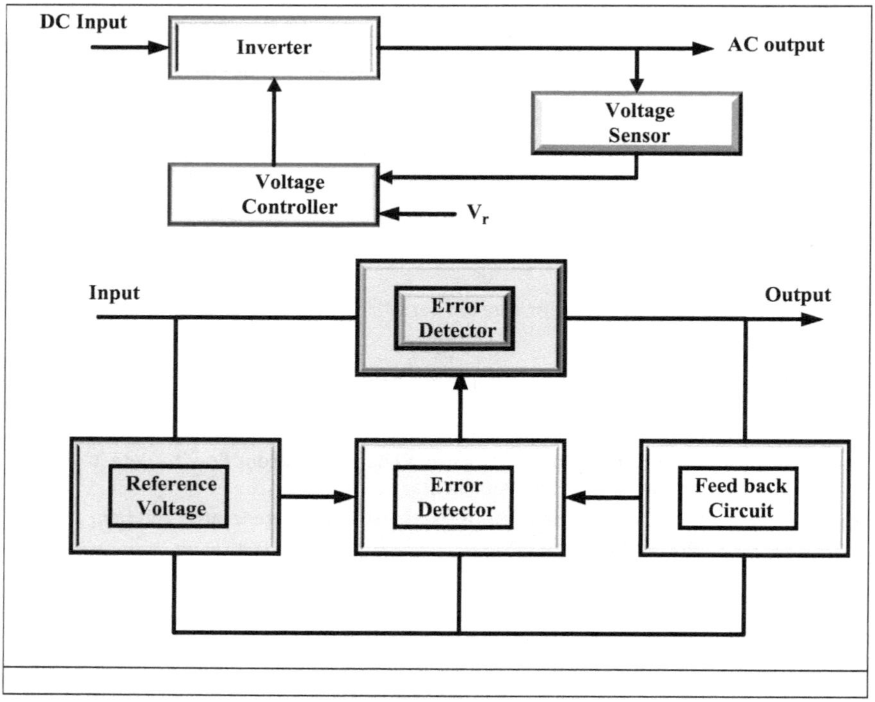

Figure 11.4 Smart voltage sensors for DFIG.

- **Wireless Connectivity:** Think about using sensors that are wireless-enabled, e.g., Bluetooth or Wi-Fi for hassle-free installation and remote monitoring. The wireless capability helps collect the data in real time and triggers proactive maintenance.
- **Integration with Monitoring Systems:** To accomplish this endeavor, the voltage sensors must be able to connect with other condition monitoring systems or SCADA systems. This integration in turn permits a centralized monitoring of the voltage data as well as a holistic way of managing the assets.
- **Data Logging and Trend Analysis:** Smart voltage sensors should have data storage ability to keep voltage trends during the long run. Many of such sensors can be connected with computer software by means of which historical voltage data can be analyzed in order to forecast equipment maintenance.
- **Self-Diagnostics:** Pay attention to the components with self-diagnostic capabilities that can evaluate their proper work. Self-diagnostics improve the reliability of voltage measurements and alert of the faults if the sensor is affected by the damage.

- **Power Efficiency:** It is advisable to consider low power sensors for the sake of prolonging battery life. Long-lasting batteries are designed to withstand the environment in which they are used, especially in areas that are hard to reach or not accessible.
- **Certifications and Standards:** Select sensors that comply with industry standards and certificates so they can be used in critical applications such as DFIGs with the guarantee of their reliability and appropriateness.
- **Durability and Environmental Resistance:** Provided the severity of the DFIG operating circumstances, pick voltage sensors that are strong, highly durable, and resistant to the external factors such as moisture, dust, and temperature variations.
- **Under and Overvoltage Protection:** Some smart voltage sensors might have built-in features like low voltage and high voltage guarding. The effects of these features are to reduce the damage to the generator unit and other connected equipment.
- **Alarm and Notification Systems:** The smart voltage sensors should be able to alert or notify the user in real time at the time of excessive voltage levels passing through the sensors. This enables the operation of the service to be proactive, stopping the occurrence of damage or downtime.

In view of the above features, operators can use smart voltage sensors which eventually enhance the reliability, safety, and efficiency of DFIG systems. Voltage measurements with high accuracy are very important for upkeeping and ensuring the health of the generator as well as avoiding electrical problems. Table 11.3 illustrates the purpose, functionality, and benefits of smart voltage sensors.

III. **Vibration Sensors:** Due to bearing failures of DFIG, a significant amount of downtime occurs in WECS. The noise of the machine changes if a fault occurs due to a back bearing fault or fore bearing fault in DFIG. To mitigate the above problem an appropriate, vibration sensor is needed.

The vibration sensors can play a crucial role in the detection of faults in DFIG. The accelerometers and piezoelectric sensors are commonly used for sensing vibration of DFIG [11–13]. Smart vibration sensors are what is referred to as the key to the monitoring of the

Table 11.3 Purpose, functionality, and benefits of smart voltage sensors

Purpose	To monitor voltage levels in the system
Functionality	These sensors measure voltage in the stator and rotor circuits for accurate control and fault detection
Benefits	These sensors Helps maintaining stable operation and provides information for diagnostics

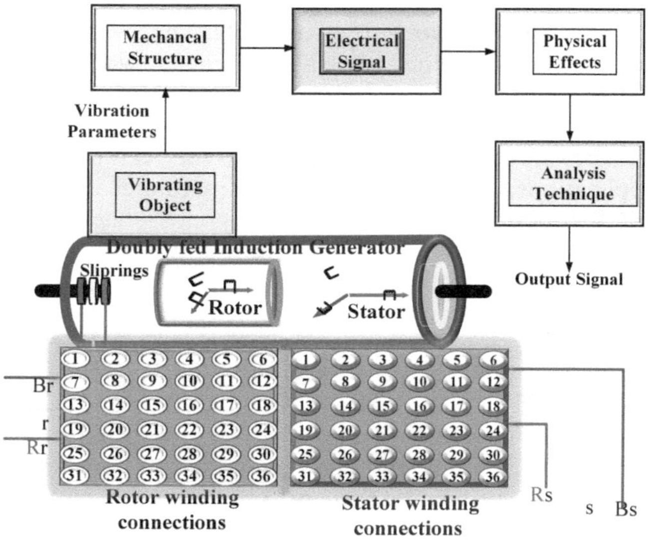

Figure 11.5 Smart vibration sensors for DFIG.

mechanical health of DFIGs. Figure 11.5 illustrates the smart vibration sensors for DFIG. These sensors are rotating machinery vibration detection devices that supply real-time data on the health of different components.

Here are some features and considerations for smart vibration sensors used in DFIG applications:

- **High Sensitivity:** A smart sensor that contains a highly vibration-sensitive sensor should be able to pick up even the slightest vibrations in the generator and system as a whole. It is therefore possible to plant an early warning flag to prevent the development of mechanical faults.
- **Frequency Range:** The sensors should cover a wide frequency range to detect vibrations connected to various faults, like, unbalance, misalignment, bearing defects, and gear problems.
- **Wireless Connectivity:** To make the installation process easy and to retrieve the data easily, think of smart sensors which are connected to the wireless network, for instance, Bluetooth or Wi-Fi. Thus, the system enables remote control and diminishes the requirement of physical presence at the generator.
- **Data Logging and Analysis:** The smart sensors must be able to record the vibration data for a long period of time. Moreover, they should have the software that is either built-in or can be easily integrated with the device to analyze the data and find the patterns and trends that are likely to be the problems.

- **Temperature Monitoring:** Some of the sophisticated vibration sensors may also have the temperature monitoring function. This gives a broader perspective of the machine's well-being, as the changes in temperature usually are associated with mechanical faults.
- **Self-Diagnostics:** The smart sensors with the self-diagnostic features have the ability to check their own functionality; hence, they will be able to collect the data that are reliable and accurate. This is the feature that gives another layer of trust in the performance of the sensor.
- **Integration with Condition Monitoring Systems:** Confirm that the vibration sensors can smoothly blend with the general condition monitoring systems or SCADA systems. This brings about the centralized monitoring of different sensors and thus a more holistic management of the assets.
- **Battery Life and Power Efficiency:** Take into account the sensors with the power consumption that is efficient so as to extend the battery life. The batteries that are designed to last for a certain period will definitely decrease the number of times the sensors need to be checked and replaced.
- **Certifications and Standards:** Choose the sensors that follow the industry standards and certifications. This way, it is certain that the sensors fulfil the quality and safety requirements for their use in these highly important applications like DFIGs.

Table 11.4 presents the purpose, functionality, and benefits of smart vibration sensors, giving a clear picture of the adaptability and the main role of these sensors to the users.

IV. **Temperature Sensors:** Smart temperature sensors are essential components for monitoring the thermal health of DFIGs. These sensors help prevent overheating and ensure the proper functioning of critical components. In the wind energy generation system, the internal temperature of DFIG is measured using sensors. These sensors are installed either on the stator periphery of the machine or rotating part of DFIG. These sensors are either thermocouples, resistance temperature detectors (RTDs), or thermistors. Moreover, the temperature of various components in the DFIG is measured by temperature sensors.

Table 11.4 Purpose, functionality, and benefits of smart vibration sensors

Purpose	To detect mechanical issues and imbalances in the rotating components
Functionality	These sensors measure vibrations in the generator and gearbox to identify potential faults
Benefits	Early detection of mechanical problems through these sensors can prevent serious damage and improve maintenance planning

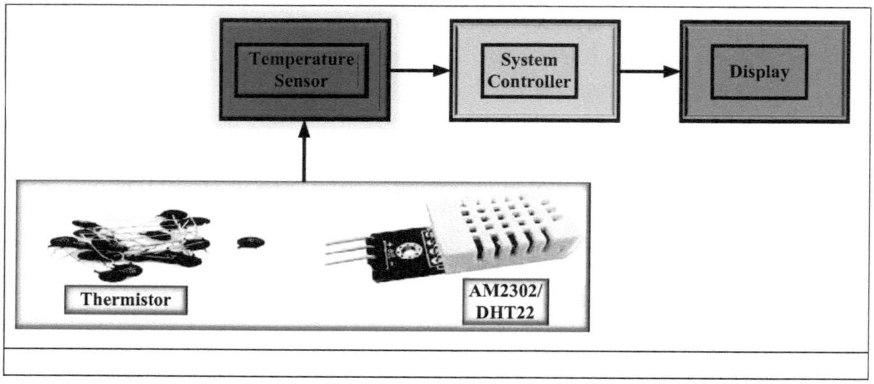

Figure 11.6 Smart temperature sensors for DFIG.

The temperature of bearings, power electronics converters, and rotor and stator windings are measured in WECS [14–16]. Figure 11.6 illustrates the smart temperature sensor for DFIG, which comprises a combination of thermocouple, RTD, thermistor, etc. When selecting smart temperature sensors for DFIG applications, consider the following features:

- **Temperature Range:** The temperature range should be chosen appropriately in sync with the temperature range of the DFIG operation. Suggest that we configure instruments to do precise measurements of the surrounding temperature, such as that of the stator windings, rotor, and bearings.
- **Accuracy and Precision:** It is not a good idea to use sensors with a not even enough accuracy and precision; it is better to use the sensors that will give precise reading related to precise temperature. It is therefore very crucial because it will help us to spot some early signs of the abnormal rise or fall in the temperature, which could be a sign of forthcoming problems.
- **Response Time:** Sensors have fast response time, and therefore, it is possible to have real-time data and ensures intervention before the overheating.
- **Wireless Connectivity:** Take into account wireless sensors that can communicate using Bluetooth, Wi-Fi, or other technologies for simple installation and remote monitoring. This way, we can get instantaneous temperature readings from the field and make preventive repair easier.
- **Integration with Monitoring Systems:** Make sure that the data from these sensors can be combined with other monitoring systems such as the SCADA systems. This integration empowers a

centralized temperature data monitoring system that allows for a total approach to asset management.

- **Data Logging and Trend Analysis:** The data loggers should capture temperature data periodically, and the sensors should have the ability to record and store temperature trends over time. Some sensors have software that analyzes historical temperature data and is programmed to send alerts, thereby helping with predictive maintenance.
- **Self-Diagnostics:** Track down the sensors that have self-diagnostic features which can assess their own functionality. The self-diagnostics of this temperature sensor allows for more reliable temperature measurements and warn of a possible malfunction of the sensor.
- **Temperature Compensation:** While choosing sensors, consider the temperature-compensating ones so that they can account for the changes in the ambient conditions. This allows the thermometer to read the accurate temperature differences under different environmental conditions.
- **Power Efficiency:** Consider choosing sensors with low power consumption so that the battery life is extended. The durability of long-lasting batteries lowers the requirement for frequent maintenance and swap, especially in areas that are difficult to access.
- **Certifications and Standards:** Select sensors that meet the industry's standards and certifications to avert the problem of nonreliability and incompatibility when used in critical applications like DFIGs.
- **Durability and Environmental Resistance:** In the case of DFIGs that usually operate under harsh conditions, the temperature sensors should be reliable, durable, and resistant to weather aspects like moisture, dust, and high temperatures.
- **Alarm and Notification Systems:** For smart temperature sensors to generate alarms or notifications promptly when the temperature threshold is surpassed, they have to be capable of doing so in real time. This permits an instant reaction to eliminate damage and avoid downtime.

Through the analysis of these characteristics, operators can select the smart temperature sensors that are a part of the DFIG system and not a problem to the system's reliability, safety, and efficiency. Early detection of temperature irregularities can help avoid overheating problems and devise maintenance planning [14–16] (Table 11.5).

V. **Wind Speed Sensors:** Smart suspended wind sensors are the key elements of wind condition monitoring for the DFIGs in wind turbines. These sensors are crucial for the turbine to perform at the highest level and to be as efficient as possible [17–20].

Table 11.5 Purpose, functionality, and benefits of temperature sensors

Purpose	To monitor the temperature of key components
Functionality	These sensors measure the temperature of the generator, gearbox, and other critical parts to prevent overheating
Benefits	Early warning of overheating through these sensors can help prevent damage and extend the lifespan of components

When selecting smart speed sensors for DFIG applications, consider the following features:

- **Measurement Range:** Pick speed sensors that would capture the desired range of rotational speeds for the DFIG. Make certain that the sensors can reliably measure both low and high speeds so it can cope with a broad range of operations.
- **Accuracy and Precision:** Use sensors with high accuracy and precision to determine the exact speed of the wind. Additional data which are generated by the speed of the generator are important for the performance of the plant and to avoid the overspeed conditions.
- **Response Time:** Instant reactions become the necessary component for catching sudden shifts on the road. Fast responsiveness of the sensors provides actual time data for immediate speed anomaly detection.
- **Wireless Connectivity:** Use sensors with wireless capability – including Bluetooth, Wi-Fi, etc. – for immediate installation and remote monitoring. The wireless connectivity lets speed data be retrieved in real time and it makes for prompt maintenance.
- **Integration with Monitoring Systems:** Provide the necessary means of integrating the speed sensors with other condition monitoring or SCADA systems. This provides the facility to track the speed data centrally and develop in all asset management a comprehensive approach.
- **Data Logging and Trend Analysis:** Data logging is an essential feature for smart speed sensors to keep track of and monitor speed patterns or trends over a period of time. Some sensors already have installed or compatible platform to analyze the speed for the past, and this helps in predictive maintenance.
- **Self-Diagnostics:** Try to find sensors that can do a self-diagnostic check to see whether they function properly. Self-diagnostics provide the reliability for the speed measurements and gives alert if the sensors fail.
- **Power Efficiency:** Search for sensors that have lower power usage to increase battery lifetime. Long-term batteries decrease the need

for maintenance tasks and spare batteries, especially in places which are difficult to access or remote.

- **Certifications and Standards:** Settle on sensors following the standard of the industrial with certifications to guarantee their dependability and suitability for critical applications like the DFIGs.
- **Durability and Environmental Resistance:** The harsh environment of DFIG operation requires the speed sensors to be strong, durable, and resistant to different environmental factors such as moisture, dust, and temperature variations.
- **Overspeed Protection:** Some smart speed sensors might have overspeed protection tools. Such functions could stop the damage when the generator is subject to an abnormal speed condition.
- **Alarm and Notification Systems:** Smart speed sensors need to be able to send alarms or notifications in real time if speed limits are surpassed. It makes it possible to take measures instantly to avert damage or downtime.

Including these features, operators can choose speed sensors that smartly contribute to the overall dependability, safety, and efficiency of DFIG systems. Correct speed measurements are fundamental in ensuring the generator's health and preventing mechanical and electrical problems. Table 11.6 presents the purpose, functionality, and benefits of smart rotational speed sensors.

This report indicates that advanced condition monitoring is universally relevant for electricity generation by wind turbines. This development can be stressed as a significant progress stage within the whole technological development of the renewable energy sector. The use of sensors, for wind turbine health monitoring and diagnosis purpose, becomes an icing on the Cake. The modern sensor technologies can be used to make the wind turbine system to be reliable and efficient. The study demonstrates the capability of smart sensors, which are primarily equipped with a data collection facility. Figure 11.7 illustrates the smart sensors surveillance parameter for DFIG. The papers on the smart sensor-based monitoring and fault diagnosis of wind turbines indicate the pivotal position of advanced sensor technologies in the implementation of wind energy systems for the highest efficacy and reliability. Such research demonstrates the benefits of smart sensors

Table 11.6 Purpose, functionality, and benefits of smart speed sensors

Purpose	To monitor the rotational speed of the generator
Functionality	These sensors measure the speed of the rotor to ensure it stays within the acceptable range
Benefits	These sensors are essential for maintaining optimal generator performance and preventing overspeed conditions

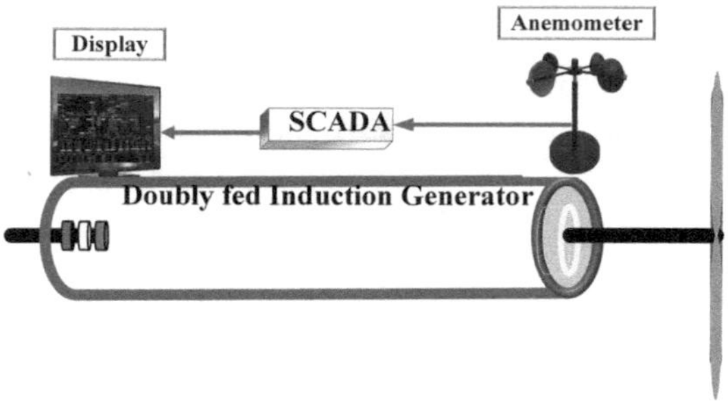

Figure 11.7 Smart speed sensors for DFIG.

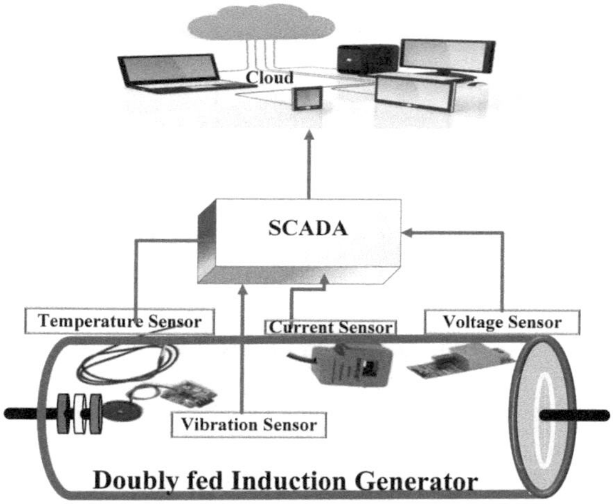

Figure 11.8 Smart sensors surveillance parameter for DFIG.

that can measure continuously critical values like temperature, vibration, and operation, which can help in early detection of any faults or abnormalities in the wind turbine machine parts (Figure 11.8).

Operators will be able to take advantage of the data generated by smart sensors integrated with data analytics and machine learning techniques to implement predictive maintenance strategies to ensure optimal turbine performance and curb downtime. Also, the smart sensors' development and application in wind turbine monitoring and control are responsible for improving the reliability of the systems and minimizing operational risks, hence the production of more energy. These studies prove the increasing role of smart sensor technologies, which helps in the development of the new state-of-the-art wind energy systems and finally moves toward more sustainable and efficient renewable energy solutions.

These digital sensors can also be integrated into the DFIG system to provide a comprehensive system for monitoring and controlling the generator. This will allow early identification of faults and enhance performance and reliability.

11.3 CONCLUSION

In this chapter, integration of smart sensors for WEC systems are presented. Smart sensors coupled with DFIG are a major leap in the field of clean energy and power generation. Such sensors, furnished with high-tech tools, have a major role of improving the surveillance, control, and maintenance of DFIGs, especially in wind turbine applications. The comprehensive set of smart sensors that are present within DFIGs caters for different aspects such as mechanical health and electrical parameters to environmental conditions. The sensors are designed to offer the comprehensive functionality that ensures the power of DFIG systems, their reliability, efficiency, and safety.

REFERENCES

1. M. Chen, L. Yu, N. S. Wade, X. Liu, Q. Liu, and F. Yang, "Investigation on the Faulty State of DFIG in a Microgrid," *IEEE Transactions on Power Electronics,* vol. 26, no. 7, pp. 1913–1919, 2011.
2. N. K. Mishra, G. Mishra, and M. K. Shukla, "6-Phase DFIG-MPPT Synergy: Pioneering Approaches for Maximizing Energy Yield in Wind Energy generation System," *International Journal of Power and Energy Conversion*, vol. 15, pp. 79–98, 2024.
3. N. K. Mishra and Z. Husain, "Application of Novel Six-phase Doubly fed Induction Generator for Open Phases through Modeling and Simulation," *IETE Journal of Research*, vol. 69, no. 6, pp. 3916–3927.
4. N. Ha, K. Xu, G. Ren, A. Mitchell, and J. Z. Ou, "Machine Learning-Enabled Smart Sensor Systems," *Advanced Intelligent Systems*, vol. 2, no. 9, pp. 1–31, 2020.
5. N. K. Mishra, G. Luthra, and M. K. Shukla, "Modeling and Simulation of Hybrid Renewable Energy System Using Real-Time Simulator," In S. Kumar, R. Setia, & K. Singh (Eds.), *Artificial Intelligence and Machine Learning in Satellite Data Processing and Services* (Vol. 970, pp. 123–135). Springer, 2023.
6. A. G. Abo-Khalil, A. Alghamdi, I. Tlili, and A. M. Eltamaly, "Current Controller Design for DFIG-Based Wind Turbines Using State Feedback Control," *IET Renewable Power Generation*, vol. 13, pp. 1938–1948, 2019.
7. M. Crescentini, S. F. Syeda, and G. P. Gibiino, "Hall-Effect Current Sensors: Principles of Operation and Implementation Techniques," *IEEE Sensors Journal*, vol. 22, no. 11, pp. 10137–10151, 2022.

8. X. Xu, J. Tang, L. He, and X. Zhong, "Design of a Novel Optical Voltage Sensor for Ultra-High Voltage Application," In *2009 Asia-Pacific Power and Energy Engineering Conference* (pp. 1–4). Wuhan, China, 2009
9. C. Wu, Y. Jiao, H. Nian, and F. Blaabjerg, "A Simplified Stator Frequency and Power Control Method of DFIG-DC System Without Stator Voltage and Current Sensors," *IEEE Transactions on Power Electronics,* vol. 35, no. 6, pp. 5562–5566, 2020.
10. J. Vidal, G. Abad, J. Arza, and S. Aurtenechea, "Single-Phase DC Crowbar Topologies for Low Voltage Ride Through Fulfillment of High-Power Doubly Fed Induction Generator-Based Wind Turbines," *IEEE Transactions on Energy Conversion,* vol. 28, no. 3, pp. 768–781, 2013.
11. S. J. Watson, B. J. Xiang, W. Yang, P. J. Tavner, and C. J. Crabtree, "Condition Monitoring of the Power Output of Wind Turbine Generators Using Wavelets," *IEEE Transactions on Energy Conversion,* vol. 25, no. 3, pp. 715–721, 2010.
12. N. K. Mishra and M. K. Shukla, "Evaluation of Various Controllers in Fractional Order Nonlinear Systems with Actuator Fault," *Multimedia Tools and Applications*, vol. 83, pp. 1–18, 2024.
13. J. Giri and N. K. Mishra, "Study on Different Fault Detection Techniques and Control Strategies in DFIG," In *Proceedings of the 2023 International Conference on Recent Advances in Electrical, Electronics & Digital Healthcare Technologies (REEDCON)*, New Delhi, India (pp. 151–156). IEEE, May 2023.
14. W. Wang et al., "Quantum Dots-Based Multiplexed Fiber-Optic Temperature Sensors," *IEEE Sensors Journal*, vol. 16, no. 8, pp. 2437–2441, 2016.
15. M. R. H. Sarker, H. Karim, R. Martinez, N. Love, and Y. Lin, "A Lithium Niobate High-Temperature Sensor for Energy System Applications," *IEEE Sensors Journal*, vol. 16, no. 15, pp. 5883–5888, 2016.
16. P. Guo, D. Infield, and X. Yang, "Wind Turbine Generator Condition-Monitoring Using Temperature Trend Analysis," *IEEE Transactions on Sustainable Energy,* vol. 3, no. 1, pp. 124–133, 2012.
17. Y. Majdoub, A. Abbou, and M. Akherraz, "Variable Speed Control of DFIG-Wind Turbine with Wind Estimation," In 2014 *International Renewable and Sustainable Energy Conference (IRSEC)* (pp. 268–274). Ouarzazate, Morocco, 2014.
18. Z. Yi, Y. Ye, M. Qin and Q.-A. Huang, "Modeling of Packaged MEMS Thermal Wind Sensor Operating on CP Mode," *IEEE Transactions on Electron Devices,* vol. 66, no. 5, pp. 2375–2381, 2019.
19. M. B. Thacker, A. Emadi, and D. A. Buchanan, "Design, Development, and Characterization of a Low Frequency CMUT-Based Anemometer," *IEEE Access*, vol. 9, pp. 127735–127741, 2021.
20. H. Bhattacherjee, D. Mukherjee, U. Vuyyuru, and C. Chakraborty, "Brushless Synchronous Generator-Unidirectional Rectifier for Offshore Wind Energy Conversion System," *IEEE Transactions on Energy Conversion,* vol. 37, no. 2, pp. 1060–1074, 2022.

Chapter 12

The potential of high-electron-mobility transistors as smart sensors

Yogesh Kumar Verma and Manoj Kumar Shukla

12.1 INTRODUCTION

The different analytical models of high-electron-mobility transistor (HEMT) are reported previously in the literature [1]. The GaN-based HEMT is recently being explored by different research groups due to the exciting properties of GaN. GaN has a wide energy bandgap of 3.4 eV at room temperature [2,3]. The wurtzite structure of GaN is thermodynamically stable in normal atmospheric conditions [4,5]. GaN exhibits higher saturation velocity and large conduction band offset [6,7–12]. The wide bandgap is an important parameter of the material to meet these needs [1].

The different performance parameters of GaN-based amplifier include voltage gain, transconductance, input impedance, output impedance, bandwidth, power gain, noise figure, output range, drain current, power dissipation, breakdown voltage, and stability [6,13]. AlGaN/GaN HEMT is a potential contender for smart sensor applications. The authors of Ref. [12] presented a simulation-based model for its application in the field of pH sensing. The parasitic capacitance needs to be modeled as these capacitances affect the input and output impedances of the amplifier. The HEMT based devices comprises of higher electron mobility due to the spatial separation of donor impurities and free electrons, due to which higher speed of operation is obtained [14,15]. HEMT-based MMICs are integral to automotive radar systems used in autonomous vehicles, where high-frequency operation and low-noise performance are critical for detecting objects and ensuring safe navigation [16,17]. HEMT is a specialized semiconductor device designed for high-frequency, high-speed applications [6,18]. It relies on the unique properties of a heterojunction between different semiconductor materials to achieve high electron mobility, which, in turn, enables its exceptional performance [4,5,7,8,19].

DOI: 10.1201/9781003633884-12

12.2 DEVICE STRUCTURE

Figure 12.1 represents the device structure. Monolithic microwave integrated circuits (MMICs) operating in the sub-terahertz and terahertz range are emerging for next-generation communication systems (6G), which will demand enormous bandwidths and very low latency. One of the key trends in MMIC development is the push for higher efficiency in power amplifiers. GaN MMICs, in particular, have shown improvements in efficiency and power handling capabilities [1,12]. By improving heat dissipation and utilizing high-efficiency load modulation techniques (e.g., Doherty amplifiers), MMICs can handle higher power with reduced thermal management issues.

HEMT device is used for different interesting applications such as low-noise amplifiers, radar systems, and communication systems [20,21]. Advances in 3D integration techniques have allowed more complex and compact MMIC designs, stacking multiple layers of circuits vertically to reduce the overall footprint. 3D MMICs are particularly useful for phased-array radar systems and multichannel communication systems. The parameters of interest for HEMT are voltage gain, input impedance, output impedance, and gain bandwidth product [22,23]. The different performance parameters of GaN-based amplifier include voltage gain, transconductance, input impedance, output impedance, bandwidth, power gain, noise figure, output range, drain current, power dissipation, break-down voltage, and stability. The AlGaN/GaN HEMT is biased in active mode and common-source configuration to analyze its performance as an amplifier. In this brief, the designing of amplifier using AlGaN/GaN

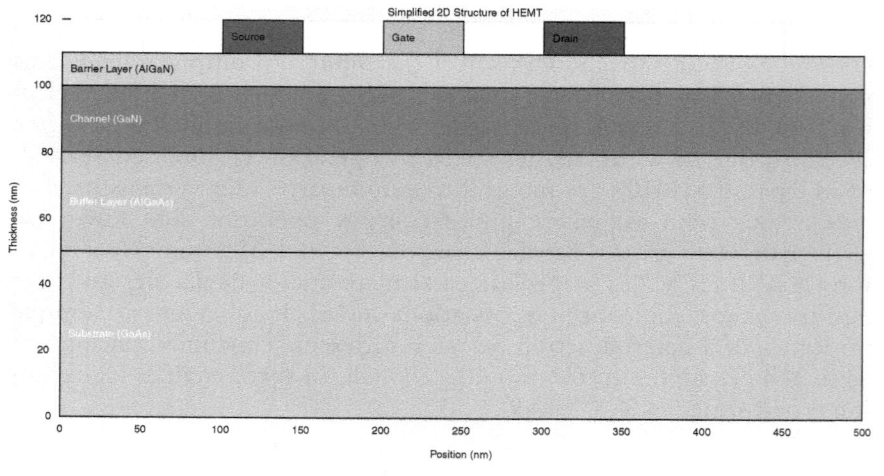

Figure 12.1 Device structure.

HEMT is performed. The frequency is varied from 10 Hz to 1 MHz to analyze the behavior of HEMT as an amplifier. The device structure is shown in Figure 12.1.

12.3 RESULTS AND DISCUSSION

Figures 12.2 and 12.3 represent the calculation of current calculated using technology-computer aided design (TCAD) for AlGaN/GaN HEMT. The different performance parameters of GaN-based amplifier include voltage gain, transconductance, input impedance, output impedance, bandwidth, power gain, noise figure, output-range, drain current, power dissipation, breakdown voltage, and stability. Figure 12.4 represents the calculation of gate-source capacitance. Figure 12.5 represents the calculation of cut-off frequency. Figure 12.6 represents the sensing characteristics of GaN-based HEMT as smart sensor.

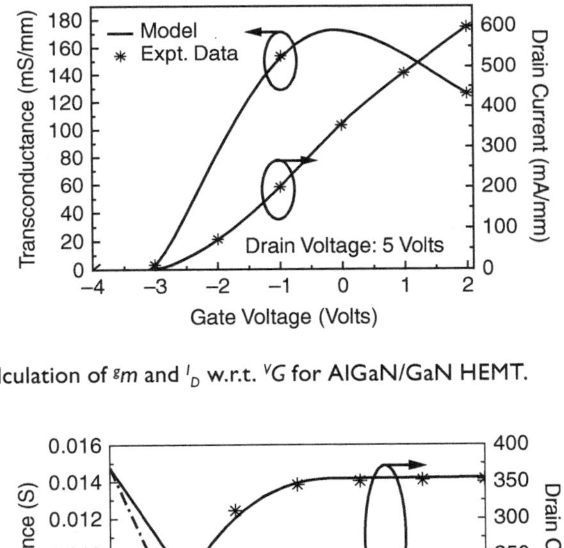

Figure 12.2 Calculation of g_m and I_D w.r.t. V_G for AlGaN/GaN HEMT.

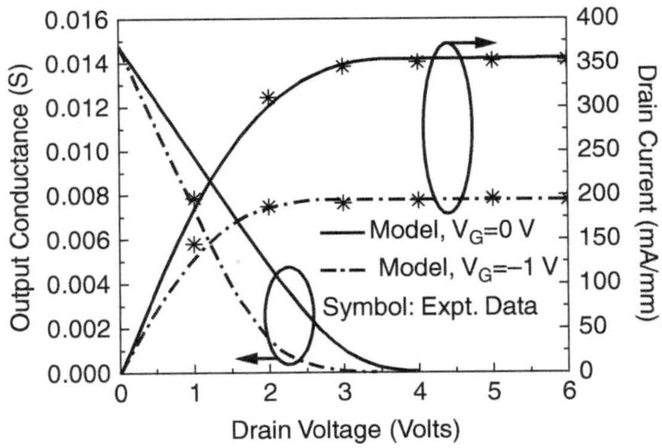

Figure 12.3 Calculation of current.

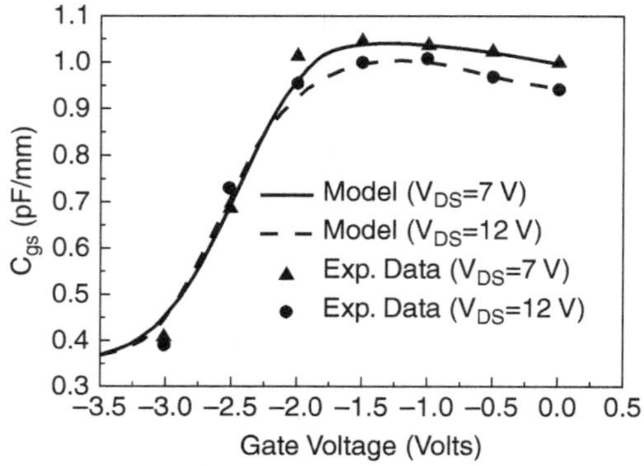

Figure 12.4 Calculation of gate-source capacitance.

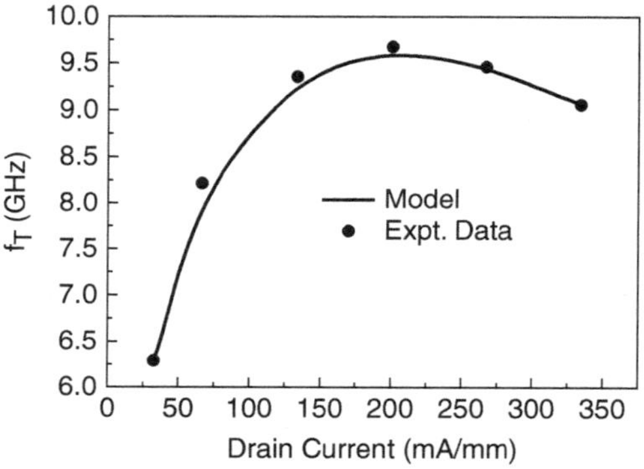

Figure 12.5 Calculation of cut-off frequency w.r.t. drain current for AlGaN/GaN HEMT.

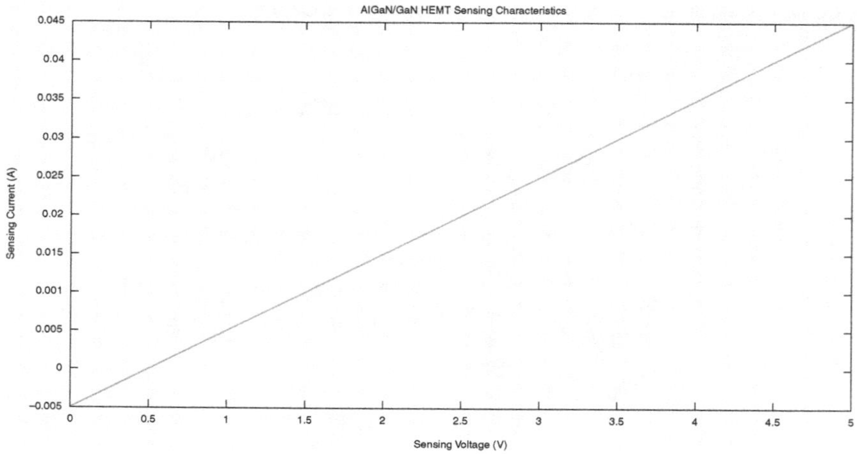

Figure 12.6 Sensing characteristics of AlGaN/GaN HEMT.

The low-noise and high-frequency performance of HEMTs make them suitable for cryogenic amplification and signal processing in quantum processors. HEMT-based MMICs are integral to automotive radar systems used in autonomous vehicles, where high-frequency operation and low-noise performance are critical for detecting objects and ensuring safe navigation. The different performance parameters of GaN-based amplifier include voltage gain, transconductance, input impedance, output impedance, bandwidth, power gain, noise figure, output-range, drain current, power dissipation, break-down voltage, and stability. Figure 12.4 represents the calculation of gate-source capacitance.

The analysis is performed at drain voltages of 7 and 12 V.

Figure 12.5 represents the calculation of cut-off frequency w.r.t. drain current for AlGaN/GaN HEMT. The peak value of cut-off frequency is obtained as 9.6 GHz at the drain current of 200 mA/mm. HEMT-based MMICs are integral to automotive radar systems used in autonomous vehicles, where high-frequency operation and low-noise performance are critical for detecting objects and ensuring safe navigation. The AlGaN/GaN HEMT provides better RF performance as compared to MOSFET (Figure 12.6).

The HEMT devices inherently comprises of polarization charges due to which there occurs the formation of 2-DEG density and therefore this device operates in depletion mode. The major benefits such as higher 2-DEG density, higher electron mobility, and high frequency operation is obtained.

12.4 CONCLUSION

In this chapter, AlGaN/GaN HEMT as a smart sensor is studied. The smart sensor based on AlGaN/GaN HEMT can be effectively used in several potential applications such as water quality testing, health monitoring systems, image processing, engine control, emission monitoring, safety systems, driver-assistance systems, waste management systems, and water management systems due to its interesting inherent features such as accuracy, reliability, and sensitivity. The applications of this device can be further extended for temperature and saline sensing, marine life sensing, and ocean life testing. Species detection and tracking are the potential applications of the proposed device. These interesting features are possible due to the use of wide-bandgap materials in the HEMT device provoking its use for higher breakdown voltages.

REFERENCES

1. M. Xu et al., "The GaAs/AlGaAs-based extended gate HEMT cardiac troponin-I biosensor: Design, mechanism and clinical detection," *IEEE Sens J*, vol. 21, no. 17, pp. 18410–18416, 2021.

2. O. Ambacher et al., "Two-dimensional electron gases induced by spontaneous and piezoelectric polarization charges in N- and Ga-face AlGaN/GaN heterostructures," *J Appl Phys*, vol. 85, no. 6, pp. 3222–3233, 1999, doi:10.1063/1.369664.
3. O. Ambacher et al., "Two-dimensional electron gases induced by spontaneous and piezoelectric polarization charges in N-and Ga-face AlGaN/GaN heterostructures," *J Appl Phys*, vol. 85, no. 6, pp. 3222–3233, 1999.
4. S. Khandelwal and T. A. Fjeldly, "Analysis of drain-current nonlinearity using surface-potential-based model in GaAs pHEMTs," *IEEE Trans Microw Theory Tech*, vol. 61, no. 9, pp. 3265–3270, 2013, doi:10.1109/TMTT.2013.2275943.
5. S. Khandelwal and T. A. Fjeldly, "A physics based compact model of I-V and C-V characteristics in AlGaN/GaN HEMT devices," *Solid State Electron*, vol. 76, pp. 60–66, 2012, doi:10.1016/j.sse.2012.05.054.
6. Y. K. Verma and S. K. Gupta, "Center Potential Based Analysis of Si and III-V Gate all around Field Effect Transistors (GAA-FETs)," *Silicon*, vol. 13, pp. 1787–1803, 2020.
7. S. Khandelwal and T. A. Fjeldly, "A physics based compact model of gate capacitance in AlGaN/GaN HEMT devices," in *2012 8th International Caribbean Conference on Devices, Circuits and Systems, ICCDCS 2012*, vol. 1, no. 1, pp. 6–9, 2012, doi:10.1109/ICCDCS.2012.6188891.
8. S. Khandelwal and T. A. Fjeldly, "Solid-state electronics a physics based compact model of I–V and C–V characteristics in AlGaN/GaN HEMT devices," *Solid State Electron*, vol. 76, pp. 60–66, 2012.
9. O. Ambacher et al., "Two-dimensional electron gases induced by spontaneous and piezoelectric polarization in undoped and doped AlGaN/GaN heterostructures," *J Appl Phys*, vol. 87, no. 1, pp. 334–344, 2000, doi:10.1063/1.371866.
10. Rashmi, A. Kranti, S. Haldar, M. Gupta, and R. S. Gupta, "Comprehensive analysis of small-signal parameters of fully strained and partially relaxed high Al-content lattice mismatched AlmGa1-mN/GaN HEMTs," *IEEE Trans Microw Theory Tech*, vol. 51, no. 2, pp. 607–617, 2003, doi:10.1109/TMTT.2002.807678.
11. M. Ga, N. Gan, H. Rashmi, A. Kranti, S. Haldar, and R. S. Gupta, "Impact of strain relaxation of Al m Ga 12m N layer on 2-DEG sheet charge density and current voltage characteristics of lattice mismatched Al," *Microelectronics*, vol. 33, no. 3, pp. 205–212, 2002.
12. A. M. Bhat, N. Shafi, C. Sahu, and C. Periasamy, "AlGaN/GaN HEMT pH sensor simulation model and its maximum transconductance considerations for improved sensitivity," *IEEE Sens J*, vol. 21, no. 18, pp. 19753–19761, 2021.
13. Y. K. Verma and S. K. Gupta, "A physics-based analytical model for ZnO based HEMT," *Micro and Nanostruct*, vol. 184, p. 207675, 2023.
14. R. Gaska, M. S. Shur, A. D. Bykhovski, A. O. Orlov, and G. L. Snider, "Electron mobility in modulation-doped AlGaN--GaN heterostructures," *Appl Phys Lett*, vol. 74, no. 2, pp. 287–289, 1999.

15. L. Yang et al., “High channel conductivity, breakdown field strength, and low current collapse in AlGaN/GaN/Si δ-Doped AlGaN/GaN: C HEMTs,” *IEEE Trans Electron Devices*, vol. 66, no. 3, pp. 1202–1207, 2019.
16. U. K. Mishra, P. Parikh, and Y.-F. Wu, “AlGaN/GaN HEMTs-an overview of device operation and applications,” *Proc IEEE*, vol. 90, no. 6, pp. 1022–1031, 2002.
17. N. Sharma et al., “High-resolution AlGaN/GaN HEMT-based electrochemical sensor for biomedical applications,” *IEEE Trans Electron Devices*, vol. 67, no. 1, pp. 289–295, 2019.
18. Y. F. Wu et al., “Bias dependent microwave performance of AlGaN/GaN MODFET's up to 100 V,” *IEEE Electron Device Lett*, vol. 18, no. 6, pp. 290–292, 1997, doi:10.1109/55.585362.
19. S. Khandelwal and T. A. Fjeldly, “A physics based compact model for drain current in AlGaN/GaN HEMT devices,” in *2012 24th International Symposium on Power Semiconductor Devices and ICs*, pp. 241–244, 2012, doi:10.1109/ISPSD.2012.6229068.
20. J.-W. Lee, V. Kumar, and I. Adesida, “High-power-density 0.25 μm gate-length AlGaN/GaN high-electron-mobility transistors on semi-insulating 6H--SiC substrates,” *Jpn J Appl Phys*, vol. 45, no. 1R, p. 13, 2006.
21. S. Arulkumaran, T. Egawa, H. Ishikawa, and T. Jimbo, “High-temperature effects of AlGaN/GaN high-electron-mobility transistors on sapphire and semi-insulating SiC substrates,” *Appl Phys Lett*, vol. 80, no. 12, pp. 2186–2188, 2002.
22. M. Nakano et al., “Electronic-field control of two-dimensional electrons in polymer-gated-oxide semiconductor heterostructures,” *Adv Materials*, vol. 22, no. 8, pp. 876–879, 2010, doi:10.1002/adma.200902162.
23. R. Singh et al., “The dawn of Ga_2O_3 HEMTs for high power electronics: A review,” *Mater Sci Semicond Process*, vol. 119, p. 105216, 2020.

Index

Kolibri 360 GmbH
Ferdinand-Jühlke-Straße 7
99095 Erfurt, Deutschland
produktsicherheit@kolibri360.de